OXFORD STUDIES IN PHYSICS

GENERAL EDITORS

B. BLEANEY, D. W. SCIAMA, D. H. WILKINSON

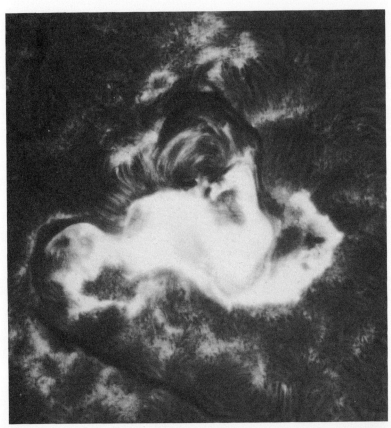

A photograph of the solar flare of 7th August 1972, from which gamma ray lines were observed. The photograph was taken in H_α radiation approximately 35 minutes after the onset of the flare. (Photograph by courtesy of Big Bear Solar Observatory, California Institute of Technology.)

GAMMA RAY ASTRONOMY

RODNEY HILLIER

Lecturer in Physics at the University of Bristol

CLARENDON PRESS · OXFORD
1984

Oxford University Press, Walton Street, Oxford OX2 6DP

London New York Toronto
Delhi Bombay Calcutta Madras Karachi
Kuala Lumpur Singapore Hong Kong Tokyo
Nairobi Dar es Salaam Cape Town
Melbourne Auckland
and associated companies in
Beirut Berlin Ibadan Mexico City Nicosia

Oxford is a trade mark of Oxford University Press

Published in the United States
by Oxford University Press, New York

British Library Cataloguing in Publication Data
Hillier, Rodney
Gamma ray astronomy.—(Oxford studies in Physics)
1. Gamma ray astronomy
I. Title
522'.686   QB471
ISBN 0-19-851451-4

Library of Congress Cataloging in Publication Data
Hillier, Rodney.
Gamma ray astronomy.
(Oxford studies in physics)
Includes bibliographies and index.
1. Gamma ray astronomy. I. Title. II. Series.
QB471.H54 1984   522'.686   83-17207
ISBN 0-19-851451-4

Filmset by Eta Services (Typesetters) Ltd, Beccles, Suffolk
Printed in Great Britain by
J. W. Arrowsmith Limited, Bristol

PREFACE

Gamma ray astronomy is slowly emerging to take its place alongside the other branches of astronomy and the contents of this book reflect a subject which is still in its pioneering stage. Whilst the potential of the subject has long been recognized, observational data are few and, in many cases, still being disputed. The presentation of the sometimes conflicting measurements posed some problems in this book; for the purpose of clarity I have not provided a complete catalogue of the published observations but instead have given examples of measurements in those fields in which real progress has been made.

The pressing need for telescopes with much greater sensitivity is obvious to everyone working in gamma ray astronomy and a large section of the book is devoted to discussion of the problems of background radiation and the design of detectors. Details of the construction of telescopes have not been included—experimentalists do not learn this craft from books—but the basic principles are discussed in a way which it is hoped will stimulate the development of new techniques and novel designs.

Throughout the book the treatment of topics starts from a level of understanding which could be expected from an undergraduate physics course. The choice of the system of units to be used is always a difficult one in a book on astronomy; undergraduate courses are now taught, almost universally, in S.I. units yet publications in astronomical journals still predominately use c.g.s. units. Not without some reservations I chose to use c.g.s. units in this book.

Whilst writing the book I was continually aware of the debts I owe to many people who have contributed to my understanding of the subject. Wherever possible these debts have been acknowledged by references in the text, but this still leaves unacknowledged the many informal contributions from lectures, discussions, and arguments. Perhaps some of my colleagues will recognize their influence in my treatment of particular topics. I should like to thank Andrew Gay who kindly read the typescript looking for errors and ambiguities; any which remain are my fault, not his. I am also grateful to Juliet Blomfield for her careful typing of several versions of the manuscript.

R.R.H.

Bristol 1983

ACKNOWLEDGEMENTS

The author acknowledges with thanks the cooperation of authors, publishers, and copyright-holders of the following figures for giving permission for their reproduction:

From *The Astrophysical Journal*, published by the University of Chicago Press, © the various authors:

3.9, p. 48; 4.6, p. 59; 5.9, p. 78; 5.10, p. 79; 7.3, p. 98; 8.18, p. 118; 9.5, p. 128; 9.6, p. 129; 9.7, p. 129; 9.8, p. 130; 9.13, p. 135; 9.14, p. 136; 9.15, p. 136; 9.16, p. 137; 9.17, p. 138; 9.18, p. 138; 9.19, p. 139; 10.1, p. 142; 10.2, p. 143; 10.3, p. 144; 10.5, p. 145; 10.9, p. 148; 10.14, p. 154; 10.17, p. 160; 10.18, p. 161; 10.19, p. 162; 10.21, p. 166; 10.22, p. 167; 10.25, p. 169; 10.26, p. 171; 11.3, p. 183; 11.4, p. 186; 11.8, p. 194; 11.9, p. 196.

From *Handbook of Physics* (London and Odishaw), published and © McGraw-Hill Book Company: 2.1, p. 5; 2.2, p. 6.

From *The atomic nucleus* (Evans), published and © McGraw-Hill Book Company: 3.1, p. 34; 3.7, p. 44; 4.1, p. 52.

CONTENTS

1

INTRODUCTION

Much was expected of gamma ray astronomy in the early 1960s. For more than twenty years it had been recognized that nuclear processes play a crucial role throughout astronomy, from the low energy nuclear reactions which generate energy in stars to the very high energy interactions of the cosmic ray particles in the interstellar medium. Gamma rays are the radiation most directly linked to nuclear processes, and measurements of gamma ray emissions can therefore provide unique information on many regions of the universe. However, gamma rays do not penetrate far into the Earth's atmosphere and it was not until the 1960s with the development of high altitude balloons and satellites that these measurements could be made. Most astronomers would identify the birth of gamma ray astronomy with the publication in 1958 of a short paper by Philip Morrison entitled 'On gamma ray astronomy' (Morrison 1958); there may have been some earlier discussion of the subject, but it was certainly this paper which provided the stimulus for the first surge of experimental work.

A search through the journals of the 1960s would show that the expectations of astronomers were not immediately realized. The early estimates, by Morrison and others, of the expected fluxes proved to be optimistic, in some cases by several orders of magnitude. This would have been less important if the fluxes originally predicted had been large, but they were generally only at the limit of detection by the early telescopes. Probably even more disappointing was the lack of any bonus in the form of an unexpectedly large flux from sources which had not been foreseen. In this respect the rapid development of X-ray astronomy in the 1960s, instead of providing a stimulus to the gamma ray work, seemed merely to emphasize its problems. Morrison's paper had made little reference to X-ray astronomy and, indeed, the first detection of a cosmic X-ray source in 1962 came quite unexpectedly during measurements of X-ray fluorescence from the moon. Nevertheless, in the course of the next few years, using telescopes of only modest size, nearly a hundred X-ray sources were discovered in the Galaxy, and this field of astronomy was quickly established.

The fundamental experimental problem which was underestimated in the early work in gamma ray astronomy was the intensity of the background radiation created in and around the detector by the energetic cosmic ray particles. Near the earth the radiation has a spectrum which is formed after many generations of interactions between cosmic ray particles and the atmosphere; this spectrum is in equilibrium with the penetrating nuclear component of the cosmic rays and it is difficult to modify it without massive shielding. The X-ray observations benefitted from the fact that this equilibrium spectrum reaches a maximum in the hard X-ray region and becomes almost negligible at lower photon energies. In retrospect it can be seen that some of the mistakes which were made in underestimating the problem of background radiation could have been avoided if it had been recognized that similar problems arise in the shielding of nuclear reactors and of neutrino experiments at particle accelerators.

The trickle of observational results in the 1960s was boosted in the 1970s by the launching of the SAS-2 and COS-B satellites carrying spark chamber telescopes. These experiments produced considerable quantities of reliable data obtained under standard conditions and established the Milky Way as a strong source of high energy gamma rays. Two of the stronger discrete sources discovered in the Galaxy were identified as the young pulsars in the Crab and Vela supernova remnants. At lower energies, gamma ray lines were observed from a solar flare in 1972, using detectors on the OSO-7 satellite. At about the same time the chance discovery of sources which produce isolated bursts of gamma rays provided welcome evidence that this subject, like the other new branches of observational astronomy, could produce surprises.

The arguments which led to the enthusiasm for gamma ray astronomy twenty years ago are still valid. But the enthusiasm must be tempered by the recognition that the real need in this subject is for improvements in experimental techniques which will lead to telescopes with increased sensitivities. It is therefore without apology that half of this book is concerned with the nature of the experimental problems and with the efforts which have been made to overcome them.

Reference

Morrison, P. (1958). *Nuovo Cimento* **7**, 858.

2

THE GENERATION OF GAMMA RAYS

2.1. Introduction

In the laboratory we generate radiation using many different mechanisms. A particular mechanism is effective over only a limited region of the electromagnetic spectrum, although this restriction is normally not inherent in the mechanism itself, but is due to limitations imposed by the techniques which are available in the laboratory. For example, we may find that we cannot produce temperatures that are sufficiently high or magnetic fields that are sufficiently strong for a particular mechanism to generate radiation at the frequency which we require. When we try to use our experience in the laboratory to understand problems in astrophysics we must keep in mind that the restrictions on these parameters will be quite different in astrophysical situations. Our knowledge of the physical conditions in stars, nebulae, and galaxies is very incomplete and to predict the gamma ray emission from these sources we shall have to draw heavily on the evidence which we have accumulated from optical astronomy, radio astronomy, and the study of cosmic rays.

In this chapter we shall consider the elementary processes which may lead to the emission of gamma rays. These can be divided into four classes:

(a) *Transitions between nuclear energy levels.* Every nucleus possesses a number of quantum energy levels and a transition from one level to another with lower energy must be accompanied by the release of energy, which may appear in the form of a gamma ray. Since the energy levels of a nucleus are discrete the gamma rays are emitted in the form of spectral lines.

(b) *The annihilation of particles with antiparticles.* The annihilation of a particle with its antiparticle results in the conversion of the rest-mass energy of the two particles into electromagnetic energy or into mesons. Of the stable elementary particles, electrons and positrons annihilate with the emission of gamma rays, whereas nucleons and antinucleons annihilate with the emission of pions.

(c) *The decays of elementary particles.* The great majority of the elementary particles are unstable and decay to more stable forms through nuclear or electromagnetic interactions. An electromagnetic decay is often associated with the emission of one or more gamma rays; for example, the neutral pion, which is produced copiously in high energy collisions between nucleons, decays with the emission of two gamma rays.

(d) *The acceleration of charged particles.* The power, P, which is radiated when a charged particle is accelerated by a force, F, is given (Jackson 1962) by

$$P = \frac{2}{3} \frac{e^2}{c^3} \left(\frac{F}{m} \right)^2 \tag{2.1}$$

where e is the charge on the particle, m is its mass, and c is the velocity of light (c.g.s. units are used throughout this book). Since the power which is radiated varies inversely as the square of the mass of the particle, in most situations we need consider only the radiation from electrons. The character of the radiation, and the name given to it, depends on the nature of the accelerating force. The radiation is *bremsstrahlung* when the electron is accelerated in the electrostatic field around a nucleus; it is *cyclotron radiation* or *synchrotron radiation* when the acceleration takes place in a static magnetic field, and the process is *Thomson scattering* or *Compton scattering* when the acceleration occurs in the electromagnetic field of a photon.

2.2. Nuclear gamma rays

2.2.1. *Radiative transitions between nuclear energy levels*

The relaxation of a nucleus from a highly excited state to the ground state can, in general, take place by several routes through intermediate energy levels and this gives rise to a large number of possible gamma ray lines. The intensities of these lines depend on the probabilities of the appropriate transitions which, in turn, are determined by the parities and the angular momenta of the states involved. The transitions fall into two categories (Cohen 1971); changes in the charge distribution in the nucleus give rise to electric multipole radiation whilst changes in the current distribution produce magnetic multipole radiation. The difference in the angular momenta of the two states, ΔJ, is carried away by the photon and the value of ΔJ determines the order of the multipole radiation, so that $\Delta J = 1$ corresponds to dipole radiation, $\Delta J = 2$ to quadrupole radiation and so on. Figures 2.1 and 2.2 show how the values of the transition probabilities, in each category, depend on the energies and the angular momenta of the photons. A transition with $\Delta J = 0$ is an example of a so-called 'forbidden' transition; although not strictly forbidden it is very slow and the nucleus may lose its energy by some other process such as particle emission.

With each isotope of the chemical elements having its own characteristic

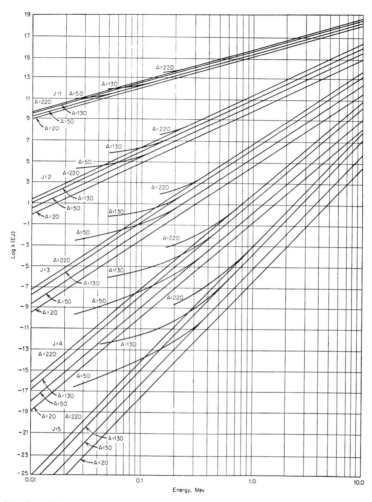

FIG. 2.1. Transition probabilities for electric transitions of a single odd proton in nuclei of mass number 20, 50, 130, and 220 for the first five multipole orders. (From Hayward 1967, p. 9–179.) © (1967) McGraw-Hill Book Company.

set of gamma ray lines the number of possible lines is very large indeed. Fortunately in most astrophysical situations we can expect the elements to be present in their natural abundances and this means that the lines of just a few elements such as carbon, oxygen, and neon should dominate the spectra; it should be noted that ^{4}He, next to ^{1}H the most abundant isotope, has no bound excited states and therefore contributes no spectral lines. Figure 2.3 shows the energy level scheme for the ^{12}C nucleus. Each level is labelled on the right-hand side with the energy in MeV and on the left-hand side with the

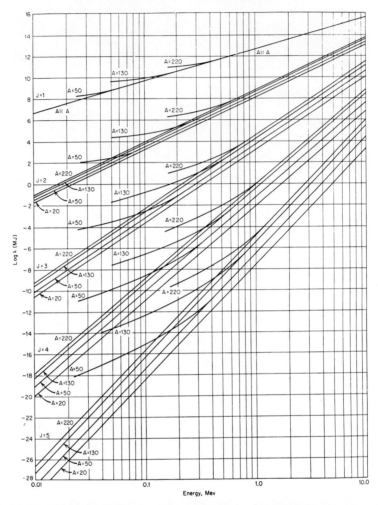

FIG. 2.2. Transition probabilities for magnetic transitions of a single odd proton in nuclei of mass number 20, 50, 130, and 220 for the first five multipole orders. (From Hayward 1967, p. 9–180.) © (1967) McGraw-Hill Book Company.

angular momentum and parity. The first excited state has angular momentum 2, and the ground state angular momentum 0, so that the transition between them results in electric quadrupole radiation of photons with an energy of 4.43 MeV. Figure 2.1 gives the lifetime of the first excited state as $\sim 10^{-12}$ s. The second excited state has the same angular momentum as the ground state so the transition between them is forbidden; it actually decays by α-emission. Figure 2.4 shows the energy level scheme for ^{16}O; in this case the transition from the first excited state is forbidden whilst that from the

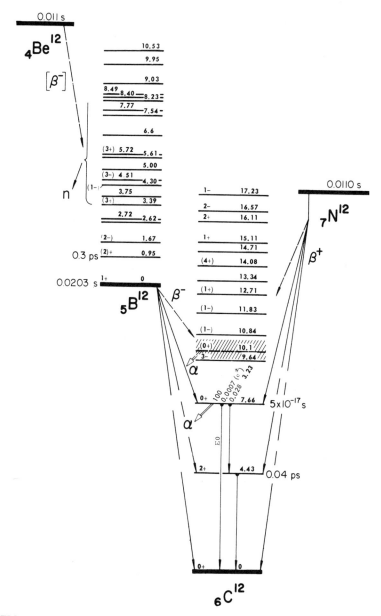

FIG. 2.3. Energy level scheme for ^{12}C and related nuclei. (From Lederer *et al.* 1968.)

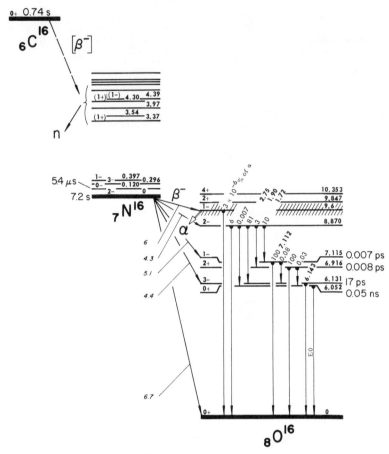

FIG. 2.4. Energy level scheme for ^{16}O and related nuclei. (From Lederer *et al.* 1968, p. 158.)

second excited state gives electric octupole radiation of photons with an energy of 6.13 MeV. Similar energy level schemes for a large number of isotopes have been given by Lederer, Hollander, and Perlman (1968).

The emission of gamma ray line spectra requires that the nuclei must first be excited. This can take place in several ways and the excitation mechanism, to some extent, determines which lines will be most prominent. Excitation processes which may be important in astrophysical situations are radioactive decay, the capture of thermal neutrons, and the scattering of fast protons and neutrons.

2.2.2. *Excitation of nuclei in collisions with fast nucleons*

The threshold energy for this form of excitation is of the order of 10 MeV;

nucleons with much greater energy may cause some target nuclei to fragment and the gamma ray spectra will then contain lines corresponding to the fragments as well as the target nuclei. The calculation of the line spectrum to be expected from this form of excitation is very difficult and much use has been made of experimental data recorded when fast protons have been fired into targets of different chemical compositions (Ramaty, Kozbovsky, and Lingernfelter 1979).

Table 2.1 gives the most prominent lines to be expected when the target composition follows the natural abundances of the elements.

Table 2.1. Gamma-ray lines excited in inter-actions with fast nucleons (from Ramaty, Kozbovsky, and Lingenfelter 1979)

Isotope	Energy of gamma-ray line (MeV)
^{12}C	4.438
^{14}N	2.313, 5.105
^{16}O	2.741, 6.129, 6.917, 7.117
^{20}Ne	1.634, 2.613, 3.34
^{24}Mg	1.369, 2.754
^{28}Si	1.779, 6.878
^{56}Fe	0.847, 1.238, 1.811

2.2.3. *Excited nuclei formed by neutron capture*

Before this process can occur the free neutron must first be produced in a separate reaction such as

$$p + {}^4He \rightarrow {}^3He + n + p.$$

Reactions of this kind are usually endothermic; in this case the incident proton must have an energy greater than 26 MeV. In material with a composition which follows the natural abundances of the elements, the neutron will normally be captured by hydrogen, forming deuterium, and releasing a gamma ray with an energy equal to the binding of the deuterium nucleus

$$n + {}^1H \rightarrow {}^2D + h\nu(2.23 \text{ MeV}).$$

The cross-section for this reaction is given by

$$\sigma_{capture} = \frac{\sigma_0}{v},$$

where v is the velocity of the neutron and $\sigma_0 = 7.3 \times 10^{-20}$ cm.s. If there are

n_H hydrogen nuclei per cm^3 the time-scale for neutron capture will be given by

$$\tau_{capture} \sim \frac{1}{n_H \sigma_{capture} v} \sim \frac{1}{n_H \sigma_0}.$$

The free neutron has a lifetime of 920 s and if capture is to occur before the neutron decays we see that

$$n_H > 1.4 \times 10^{16} \text{ nuclei cm}^{-3}.$$

Thus free neutrons produced deep inside stars will undergo capture whilst neutrons produced in the interstellar medium will, in general, decay before they are captured.

The cross-section for the elastic scattering of neutrons in hydrogen is $\sim 10^{-23}$ cm^2, which is very much larger than the cross-section for capture, except at very low neutron velocities. The neutrons therefore lose their energy very quickly in elastic collisions and are captured when they have only thermal energies.

2.2.4. *Excited nuclei formed in radioactive decay*

In this case the gamma rays come from a daughter nucleus which has been formed in an excited state. If, initially, there are n_0 parent nuclei present, with mean lifetime τ, the number remaining after a time t is given by

$$n(t, \tau) = n_0 \exp(-t/\tau).$$

The activity, $A(t, \tau)$, of the nucleus is defined by

$$A(t, \tau) = \frac{dn(t, \tau)}{dt} = \frac{n_0}{\tau} \exp(-t/\tau).$$

In the natural abundance of elements the only radioactive species present are those with very long lifetimes and which therefore have very low activities. Radioactive nuclei with much shorter lifetimes may be synthesized in the interiors of some stars, but since they decay quickly the nuclei must be brought rapidly to the surface of the stars if the gamma rays produced in the decays are not to be absorbed in the stellar material.

If a number of different radioactive isotopes, each with its characteristic lifetime, are produced in equal abundances then the isotope with the greatest activity at some time, t, later will be that for which

$$\frac{dA(t, \tau)}{d\tau} = 0;$$

that is, the isotope with

$$\tau = t.$$

Arnett and Clayton (1970) have considered the production of radioactive nuclei in a model of a supernova explosion in which large quantities of ^{56}Ni are formed from the fusion of ^{28}Si. The ^{56}Ni decays to ^{56}Fe in two β-decay processes,

$$^{56}Ni + e^- \xrightarrow{\tau=6.1d} {}^{56}Co^* + \nu$$

$$\xrightarrow{\tau=77d} {}^{56}Fe^* + e^+ + \nu.$$

The energy level schemes for ^{56}Ni, ^{56}Co, and ^{56}Fe are shown in Fig. 2.5. Arnett and Clayton conclude that if observations are made at a period of the order of a month after the explosion the most prominent gamma ray lines will be the 0.812 MeV line from ^{56}Ni and the 0.847 MeV line from ^{56}Co. The annihilation radiation from the positrons will be considered in the next section.

The Crab nebula is a supernova remnant which is nearly a thousand years old and its activity is probably dominated by ^{249}Cf which has a half-life of 330 y and gives strong gamma ray lines at 0.34 MeV and 0.39 MeV.

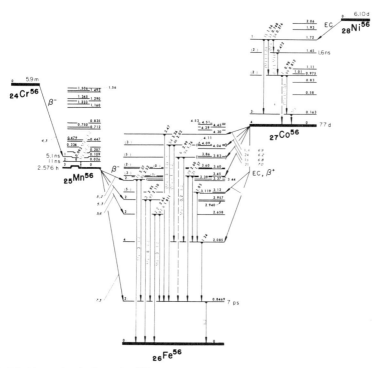

FIG. 2.5. Energy level scheme for ^{56}Fe and related nuclei. (From Lederer *et al.* 1968, p. 189.)

2.2.5. *Annihilation of positrons with electrons*

When a positron annihilates with an electron two or more gamma rays are produced. The interaction where both the positron and the electron are at rest, and where only two gamma rays emerge, is of particular interest because each gamma ray then has the characteristic energy

$$\varepsilon = \tfrac{1}{2}m_ec^2 = 0.511 \text{ MeV}.$$

This line spectrum of radiation is referred to as *annihilation radiation*.

The electron and the positron both have spins of $\tfrac{1}{2}$ and at low energies they may be in either of two possible angular momentum states, namely a $^1S_0$ state with $J = 0$ or a $^3S_1$ state with $J = 1$. If the initial spins of the two particles are randomly orientated the $^3S_1$ state occurs three times as frequently as the $^1S_0$ state. Selection rules forbid the decay of the $^1S_0$ state into an odd number of gamma rays and the decay of the $^3S_1$ state into an even number of gamma rays. In each case, the decay with the least number of photons is the most probable, so the $^1S_0$ state usually decays into two gamma rays and the $^3S_1$ state usually decays into three gamma rays.

The cross-section for the two-photon decay is given (Heitler 1954) by

$$\sigma_{2\gamma} = \frac{\pi r_0^2 c}{v} \tag{2.2}$$

where v is the velocity of the positron relative to the electron and

$$r_0 = \frac{e^2}{m_ec^2}.$$

The cross-section for the three-photon decay is given (Heitler 1954) by

$$\sigma_{3\gamma} = \tfrac{4}{3}(\pi^2 - 9)\alpha \frac{r_0^2 c}{v}, \tag{2.3}$$

where

$$\alpha = \frac{e^2}{\hbar c}.$$

From eqns 2.2 and 2.3 we see that

$$\frac{\sigma_{2\gamma}}{\sigma_{3\gamma}} \sim 372. \tag{2.4}$$

If the relative velocity of the electron and the positron is very small they may form a bound system before annihilating. This bound system, known as *positronium*, is analogous to a hydrogen atom. The relative frequency of the two-photon and the three-photon decay of positronium differs from that given by eqn 2.4 because, once positronium is formed in a definite angular momentum state, it remains in that state until annihilation occurs.

Statistically the $^3S_1$ state is formed three times as frequently as the $^1S_0$ state so the relative frequency of the two decay modes of positronium is

$$\left(\frac{\sigma_{2\gamma}}{\sigma_{3\gamma}}\right)_{\text{positronium}} = \frac{1}{3}.$$

The two-photon decay produces a spectral line at $\varepsilon = 0.511$ MeV. The natural width of this line, which is given by

$$\Delta\varepsilon \sim \frac{\hbar}{\tau},$$

is only 5×10^{-6} eV.

2.3. The decay of neutral pions

2.3.1. *The kinematics of the decay of neutral pions*

The neutral pion is unstable and decays with a lifetime of $\sim 10^{-16}$ s into two gamma rays:

$$\pi^0 \rightarrow h\nu + h\nu.$$

When the pion decays at rest each gamma ray has an energy of $\frac{1}{2}m_\pi c^2$, where m_π is the mass of the pion. But neutral pions are seldom created at rest and since they suffer no energy loss by ionization they usually decay in flight.

Consider the decay of a neutral pion which has a velocity $\beta_\pi c$ in the observer's reference frame. In the reference frame of the pion the two gamma rays will be emitted in opposite directions, each with energy ε_0^* where $\varepsilon_0^* = \frac{1}{2}m_\pi c^2$. Let the angles between the directions of the gamma rays and the velocity of the pion be α^* and $(\pi - \alpha^*)$, respectively. Since the pion has zero spin the angular distribution of the gamma rays will be isotropic in the pion reference frame and the number of gamma rays emitted between α^* and $\alpha^* + d\alpha^*$ will be given by

$$n(\alpha^*)\, d\alpha^* = \sin \alpha^* \, d\alpha^*.$$

A gamma ray which is emitted at an angle α^* in the pion reference frame will have an energy ε in the observer's reference frame where

$$\varepsilon = \frac{\varepsilon_0^*(1 + \beta_\pi \cos \alpha^*)}{\sqrt{(1 - \beta_\pi^2)}}. \tag{2.5}$$

The energy distribution of the gamma rays in the observer's reference frame will be given by

$$n(\varepsilon)\, d\varepsilon = n(\alpha^*)\, d\alpha^*,$$

i.e.

$$n(\varepsilon) = \frac{1}{\varepsilon_C^*}\frac{\sqrt{(1 - \beta_\pi^2)}}{\beta_\pi}. \tag{2.6}$$

We see from eqns 2.5 and 2.6 that the energy distribution of the gamma rays is flat and extends from

$$\varepsilon_{\min} = \varepsilon_0^* \sqrt{\{(1 - \beta_\pi)/(1 + \beta_\pi)\}} \tag{2.7}$$

to

$$\varepsilon_{\max} = \varepsilon_0^* \sqrt{\{(1 + \beta_\pi)/(1 - \beta_\pi)\}} . \tag{2.8}$$

On a logarithmic scale the energy distribution is symmetric about $\varepsilon = \varepsilon_0^*$. Therefore the energy distribution of the gamma rays resulting from the decay of many pions, with different values of β_π, can be obtained by adding many rectangles each centred on $\varepsilon = \varepsilon_0^*$ and with dimensions given by eqns 2.6, 2.7, and 2.8. We see that the spectrum always has a maximum at $\varepsilon = \varepsilon_0^*$ and, when plotted on a logarithmic energy scale, is symmetric about this energy.

Neutral pions may be created through a variety of interactions, the most important in astrophysical problems being collisions between high energy nucleons and the annihilation of nucleons with antinucleons.

2.3.2. *The production of neutral pions in collisions between high energy nucleons*

The number of baryons must be conserved in a collision and, in the case of a collision between two protons, the simplest interaction involving the production of pions is

$$p + p \rightarrow p + p + \pi^0.$$

We shall first calculate the kinetic energy which is available in the centre-of-momentum system (c.m.s.) of the two protons because this must be greater than the rest-mass energy of the neutral pion if the reaction is to proceed.

Consider a collision between a proton with mass M, momentum P_1, total energy E_1, and a stationary proton. The momenta of the two protons in the c.m.s. system will be equal, so we may write

$$\frac{P_1 - \beta_c E_1/c}{\sqrt{(1 - \beta_c^2)}} = \frac{M\beta_c c}{\sqrt{(1 - \beta_c^2)}},$$

where β_c is the velocity of the c.m.s. system in the observer's reference frame. Hence,

$$\beta_c = \frac{cP_1}{E_1 + Mc^2} = \sqrt{\{(E_1 - Mc^2)/(E_1 + Mc^2)\}} .$$

The energy of each proton in the c.m.s. system will be given by

$$E^* = \frac{Mc^2}{\sqrt{(1 - \beta_c^2)}} .$$

There will be sufficient energy to create a pion with mass m_π, if

$$\frac{2Mc^2}{\sqrt{(1 - \beta_c^2)}} > (2M + m_\pi)c^2,$$

i.e.
$$\beta_c > \sqrt{\{1 - (1 + m_\pi/2M)^{-2}\}}. \qquad (2.9)$$

From eqn 2.9 we see that the threshold kinetic energy of the incident proton is given by

$$T_1 = E_1 - Mc^2 = 2m_\pi c^2(1 + m_\pi/4M).$$

Numerically, $T_1 = 296$ MeV.

At threshold the pion will be at rest in the c.m.s. system but in the observer's system it will have a velocity

$$\beta_\pi = \sqrt{\{1 - (1 + m_\pi/2M)^{-2}\}}.$$

When the incident proton collides, not with another proton, but with a complex nucleus the pion is still created in a single proton–nucleon collision; however, in this case, the target nucleon is not at rest but has the Fermi momentum which is characteristic of a nucleon in a nucleus. A typical value of the Fermi momentum is ~ 200 MeV/c and if a target nucleon with this momentum happens to be moving towards the incident proton the threshold energy is reduced from 296 MeV to ~ 170 MeV.

When the incident proton has an energy much greater than the threshold value, more than one pion may be created in the collision and the pions may not be at rest in the c.m.s. system. There is no complete theory of pion production in high energy collisions but the experimental results can be expressed in the form of two parameters; these parameters are the inelasticity, η, which is the fraction of the initial energy which is converted into pions and the multiplicity, v, which is the number of pions created.

Let us write

$$\eta(E) \propto E^\delta$$

and
$$v(E) \propto E^\alpha.$$

If the incident protons have an energy spectrum of the form

$$n(E) \, dE = n_0 E^{-m} \, dE,$$

then the pions created in the collisions will have a spectrum

$$n_\pi(E_\pi) \, dE_\pi \propto n(E)v(E) \, dE,$$

where
$$E_\pi \propto E \frac{\eta(E)}{v(E)}.$$

We find
$$n_\pi(E_\pi)\, dE_\pi \propto E_\pi^{-p}\, dE_\pi$$
where
$$p = (m + \delta - 2\alpha)/(1 + \delta - \alpha).$$

Experimentally it is found that
$$\delta \sim 0$$
and
$$\alpha \sim 0.25$$
so that
$$p \sim \tfrac{4}{3}(m - \tfrac{1}{2}).$$

The spectrum of the gamma rays which result from the decay of these pions can be calculated by noting that the expression given in eqn 2.6 for the gamma ray spectrum from pions with a velocity $\beta_\pi c$,
$$n(\varepsilon)\, d\varepsilon = \frac{1}{\varepsilon_0} \frac{\sqrt{(1 - \beta_\pi^2)}}{\beta_\pi}\, d\varepsilon,$$
can be written as
$$n(\varepsilon)\, d\varepsilon = \frac{1}{\sqrt{(E_\pi^2 - m_\pi^2 c^4)}}\, d\varepsilon.$$

The gamma ray spectrum from the decay of pions with a distribution of energies is then
$$n(\varepsilon)\, d\varepsilon = \int_{E_{min}}^{\infty} \frac{n_\pi(E_\pi)\, dE_\pi}{\sqrt{(E_\pi^2 - m_\pi^2 c^4)}}.$$

The lower limit to the integration occurs because, as we have seen in eqn 2.7, pions with velocities below β_{min}, where
$$\beta_{min} = (\varepsilon_0^{*2} - \varepsilon^2)/(\varepsilon_0^{*2} + \varepsilon_0^2),$$
cannot contribute gamma rays with energy ε. Thus the value of E_{min} is given by
$$E_{min} = \varepsilon + m_\pi^2/4\varepsilon.$$

2.3.3. *The production of pions in the annihilation of nucleons with antinucleons*

Antiprotons and antineutrons may be created in collisions between high energy nucleons but, at moderate energies the ratio of antinucleons to mesons is small and the subsequent annihilation of the antinucleons is not a major source of gamma rays. However, cosmologically, the existence of large regions of antimatter cannot be excluded and, in the boundary regions between matter and antimatter, annihilation rates may be very large.

The cross-section for annihilation may be written as

$$\sigma_a = \sigma_0/\beta,$$

where βc is the velocity of either particle in the c.m.s. system and $\sigma_0 \sim 2.4 \times 10^{-26}$ cm^2 (Stecker 1971).

The annihilation of a proton with an antiproton results in the emission of a number of pions,

$$p + \bar{p} \to n\pi.$$

Selection rules preclude values of n less than 3 and energy considerations determine that

$$n \leqslant 2M/m_\pi < 14.$$

The maximum energy of a pion, if annihilation takes place between a proton and an antiproton at rest, occurs when $n = 3$ and is given by

$$E_\pi = (4M^2 - 3m_\pi)/4M \sim 923 \text{ MeV}.$$

Thus, from eqns 2.7 and 2.8, the gamma ray spectrum from annihilation at rest must lie between 5 MeV and 919 MeV. The form of the spectrum between these limits has been calculated by Stecker (1971) and is shown in Fig. 2.6.

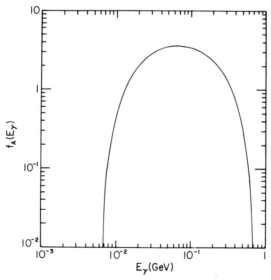

FIG. 2.6. Normalized spectrum of gamma rays from the decay of pions produced when protons and antiprotons annihilate at rest. (From Stecker 1971, p. 188.)

2.4. Synchrotron radiation

A non-relativistic electron moving with velocity $\mathbf{v}$ in a magnetic field $\mathbf{H}$ experiences the Lorentz force, $\mathbf{F}$, where

$$\mathbf{F} = \frac{e}{c}(\mathbf{v} \times \mathbf{H}). \tag{2.10}$$

As a consequence of this force the electron is constrained to move in a helix about the direction of the magnetic field. The time, T, taken by the electron to make one revolution of the magnetic field is given by

$$T = 2\pi mc/eH.$$

The frequency, ν_L, where

$$\nu_L = 1/T = eH/mc,$$

is known as the Larmor frequency. The acceleration suffered by the electron causes it to radiate, and when the electron is non-relativistic, the radiation is known as cyclotron radiation. From eqns 2.1 and 2.10 the power which is radiated is given by

$$P(E, H) = \frac{2e^4 H_\perp^2 v^2}{3m^2 c^3} \tag{2.11}$$

where $H_\perp$ is the component of the magnetic field perpendicular to the velocity. The emission occurs mainly at a frequency equal to the Larmor frequency. Seen from a direction parallel to the magnetic field the radiation is circularly polarized whilst seen from a direction perpendicular to the magnetic field it is plane polarized.

Synchrotron radiation is the name given to the radiation emitted by relativistic electrons moving in a magnetic field and it differs in several respects from cyclotron radiation. The Larmor frequency, ν'_L, of a relativistic electron is given by

$$\nu'_L = \frac{eH_\perp}{2\pi mc} \frac{mc^2}{E}.$$

The polar diagram of the electromagnetic field around a relativistic electron is shown in Fig. 2.7. The field is concentrated about the direction of motion of the electron and the angular width, θ, of the forward lobe of the polar diagram is given by

$$\theta \sim mc^2/E.$$

Thus a distant observer, O', sees bursts of electromagnetic field as shown in Fig. 2.8. The interval, T', between the bursts is given by

$$T' = 1/\nu'_L$$

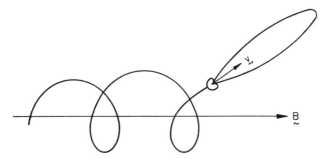

FIG. 2.7. Polar diagram of the electromagnetic field of a relativistic electron moving in a magnetic field. (From Tucker 1975, p. 41.)

and the duration, $\Delta T'$, of each burst is given by

$$\Delta T' \sim \frac{\theta(1 - v/c)}{2\pi v'_L}. \tag{2.12}$$

The factor $(1 - v/c)$ represents the shortening of the burst, as seen by the observer, due to the fact that the electron is moving in the same direction as its radiation. Now, when $v \sim c$

$$(1 - v/c) \sim (E/mc^2)^2,$$

so eqn 2.12 becomes

$$\Delta T' \sim \frac{eH}{mc}\left(\frac{E}{mc^2}\right)^2.$$

The spectrum of the synchrotron radiation can be obtained by taking the Fourier transform of the waveform in Fig. 2.8. The spectrum consists of very

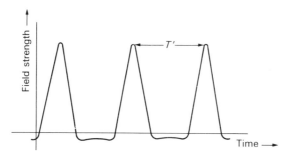

FIG. 2.8. Electromagnetic field of the electron in Fig. 2.7, as measured by a distant observer.

closely spaced harmonics of the relativistic Larmor frequency, v_L', up to a maximum frequency, v_m, where

$$v_m \sim \frac{1}{2\pi\,\Delta T'} \sim \frac{eH_\perp}{2\pi mc}\left(\frac{E}{mc^2}\right)^2.$$

The envelope of these harmonics is shown in Fig. 2.9 and it has a maximum at a frequency, v_p, where $v_p \sim 0.44\,v_m$. If we measure v_p in Hz, $H_\perp$ in gauss, and E in MeV we find that

$$v_p \sim 4.6 \times 10^6\,HE^2. \tag{2.13}$$

Clearly, only extremely energetic electrons moving in very strong magnetic fields will generate synchrotron radiation at gamma ray frequencies. But the importance of synchrotron radiation to gamma ray astronomy is that, when it is observed in other regions of the electromagnetic spectrum, it indicates the existence of relativistic electrons which are capable of producing gamma rays through other interactions such as bremsstrahlung or Compton scattering.

The expression for the power radiated by a relativistic electron must be invariant under a Lorentz transformation (since energy and time transform in a similar way) and the expression must also reduce to that given in eqn 2.11 when the electron is non-relativistic. Using these arguments Bless (1968) has shown that the total power radiated is given by

$$P(E, H) = \frac{2e^4 H_\perp^2}{3m^2 c^3}\left(\frac{E}{mc^2}\right)^2. \tag{2.14}$$

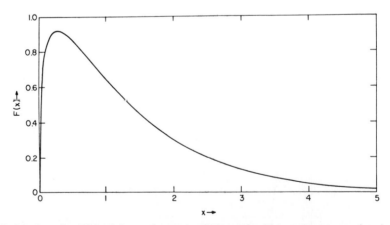

FIG. 2.9. Intensity, $F(x)$, of the synchrotron radiation emitted by an electron as a function of the parameter, x, where $x = 4\pi m^3 c^5 v / 3eHE^2$. (From Tucker 1975, p. 120.)

$H_\perp$ is the component of the magnetic field in a direction perpendicular to the velocity of the electron; if we have a large number of electrons with their velocities distributed isotropically, we can write

$$\langle H_\perp^2 \rangle = \tfrac{2}{3}H^2. \tag{2.15}$$

The radiation from an individual electron is elliptically polarized but the radiation from an isotropic distribution of electrons is plane polarized, the plane of polarization being perpendicular to the direction of the magnetic field.

In many astrophysical problems the radiation is produced by electrons with a wide range of energies. Let $n(E)\,dE$ be the number of electrons with energies between E and $E + dE$; then the spectrum of the radiation produced by these electrons will be

$$J(v)\,dv = \int_E n(E)P(E, H)\,dE\,dv. \tag{2.16}$$

Fortunately, many problems require only an approximate solution to eqn 2.16. In such cases we may assume that all the radiation from an electron is emitted at a single frequency, v_p. Equation 2.16 then becomes

$$J(v)\,dv = \int_E n(E)P(E, H)\,\delta(v - v_p)\,dv. \tag{2.17}$$

It should be remembered that the frequency, v_p, is a function of E, the electron energy.

One example of a problem which occurs frequently in astrophysics and which can be solved using the approximate solution in eqn 2.17 is the calculation of the spectrum which is produced from electrons with an energy spectrum in the form of a power law. In this case we may write

$$n(E)\,dE = n_0 E^{-m}\,dE.$$

Substituting this into eqn 2.17, and using the expression for $P(E, H)$ given in eqn 2.14, we find

$$J(v) = J_0 v^{-(m-1/2)}, \tag{2.18}$$

where

$$J_0 = \frac{n_0 K(kH)^{m+1/2}}{2},$$

$$K = \frac{2e^3}{3m^3c^5},$$

and

$$k = \frac{eH}{2\pi m^3 c^5}.$$

Thus the frequency spectrum of the synchrotron radiation, like the energy spectrum of the electrons, has the form of a power law,

$$J(v) \propto v^{-\alpha},$$

where

$$\alpha = \frac{m-1}{2}. \tag{2.19}$$

2.5. Compton scattering

In nuclear physics the interaction between a stationary electron and a photon with energy much less than mc^2 (where m is the mass of an electron) is known as Thomson scattering, whilst the interaction between an electron and a photon with energy greater than mc^2 is known as Compton scattering. It is customary in astrophysics to refer to both processes as Compton scattering. In this chapter, where we derive the basic formulae associated with the interactions, we shall differentiate between the two cases but in later chapters we shall refer to both as Compton scattering.

In the laboratory Thomson and Compton scattering are usually seen in a reference frame in which the electron is initially at rest. The cross-section, σ_T, for Thomson scattering is (Heitler 1954)

$$\sigma_T = \frac{8\pi r_0^2}{3}, \tag{2.20}$$

where r_0 is the classical radius of the electron,

i.e.
$$r_0 = \frac{e^2}{mc^2}.$$

The photon loses very little energy in the interaction and the angular distribution of the scattered radiation is given by

$$n(\theta)\,d\Omega = \frac{\pi r_0^2}{2}(1 + \cos^2\theta)\,d\Omega, \tag{2.21}$$

where θ is the angle through which the photon is scattered and

$$d\Omega = 2\pi \sin\theta\,d\theta.$$

In astrophysics the interaction is often seen in a reference frame in which the electron is initially not at rest, but is moving with a relativistic velocity, and in these circumstances many photons gain energy as a result of the interaction.

Consider collisions between an electron with velocity βc and a photon with energy ε_0. If we assume that the initial direction of motion of the photon is distributed isotropically then the average energy of the photon in the electron's reference frame will be

$$\langle \varepsilon_0^* \rangle = \frac{1}{\sqrt{(1 - \beta^2)}} \varepsilon_0 ,$$

or
$$\langle \varepsilon_0^* \rangle = \frac{E}{mc^2} \varepsilon_0 \qquad (2.22)$$

where E is the total energy of the electron. Thus the condition that the interaction can be interpreted as Thomson scattering is

$$\langle \varepsilon_0^* \rangle \ll mc^2 ,$$

or
$$E\varepsilon_0 \ll m^2 c^4 .$$

The energy distribution of the scattered photons can be derived by combining eqn 2.21 with eqn 2.20, and then transforming back into the observer's reference frame. The distribution which is obtained (Ginzburg 1969) is

$$s(E, \varepsilon) = \frac{\pi r_0^2 m^4 c^8}{4 \varepsilon_0^2 E^3} \left\{ \frac{2\varepsilon}{E} - \frac{\varepsilon m^2 c^4}{\varepsilon_0 E^3} + \frac{4\varepsilon}{E} \ln \left(\frac{\varepsilon m^2 c^4}{4 \varepsilon_0 E^2} \right) + \frac{8 \varepsilon_0 E}{m^2 c^4} \right\}. \qquad (2.23)$$

The average energy, $\langle \varepsilon \rangle$, of the scattered photon is

$$\langle \varepsilon \rangle = \tfrac{4}{3} \varepsilon_0 \left(\frac{E}{mc^2} \right)^2 . \qquad (2.24)$$

Equation 2.24 can be derived, apart from a small numerical factor, by a very elementary calculation. Consider a head-on collision between the photon and the electron. In this case we have

$$\varepsilon_0^* = \varepsilon_0 \sqrt{\left(\frac{1 + \beta}{1 - \beta} \right)}. \qquad (2.25)$$

We shall consider just two extreme values for the scatering angle, θ^*. If $\theta^* = 0$, the direction of the photon is unchanged by the interaction and when we transform back into the observer's reference frame we find

$$\varepsilon = \varepsilon^* \sqrt{\left(\frac{1 - \beta}{1 + \beta} \right)} \sim \frac{\varepsilon^*}{2} \sqrt{(1 - \beta^2)}. \qquad (2.26)$$

From eqns 2.25 and 2.26 we see

$$\varepsilon = \varepsilon_0 . \qquad (2.27)$$

If $\theta^* = \pi$, the direction of motion of the photon is reversed in the interaction and when we transform back into the observer's reference frame we find

$$\varepsilon = \varepsilon^* \sqrt{\left(\frac{1+\beta}{1-\beta}\right)}$$

i.e.
$$\varepsilon \sim 4\varepsilon_0 \left(\frac{E}{mc^2}\right)^2. \qquad (2.28)$$

To a first approximation the average energy of the scattered photon is the mean of the two values given in eqns 2.27 and 2.28

i.e.
$$\langle \varepsilon \rangle \sim 2\varepsilon_0 \left(\frac{E}{mc^2}\right)^2. \qquad (2.29)$$

We see that eqns 2.24 and 2.29 are similar except for the numerical factor.

We shall now consider the power which is lost by an electron moving in a region where photons with a range of energies are present. Let $r(\varepsilon)\,d\varepsilon$ represent the density of photons with energies between ε and $\varepsilon + d\varepsilon$. Since the Thomson cross-section is independent of energy we may write

$$dP(E, \varepsilon) = r(\varepsilon)\sigma_T c\tfrac{4}{3} \left(\frac{E}{mc^2}\right)^2 \varepsilon\,d\varepsilon.$$

Integrating, we obtain

$$P(E, u) = \tfrac{4}{3}\sigma_T c \left(\frac{E}{mc^2}\right)^2 u,$$

where u, the total energy density in the radiation, is given by

$$u = \int_\varepsilon r(\varepsilon)\varepsilon\,d\varepsilon.$$

If we write σ_T in terms of atomic constants we obtain

$$P_T(E, u) = \frac{32\pi e^4 u}{9m^2 c^3} \left(\frac{E}{mc^2}\right)^2. \qquad (2.30)$$

We may compare this with eqn 2.14, which gives the power radiated in the form of synchrotron radiation

$$P_S(E, H) = \frac{32\pi e^4}{9m^2 c^3} \frac{H^2}{8\pi} \left(\frac{E}{mc^2}\right), \qquad (2.30a)$$

where we have used the relationship between H and $H_\perp$ given in eqn 2.15. Equations 2.30 and 2.30a are similar, the distinctive factor in each equation being the energy density in the electromagnetic field.

As in the case of synchrotron radiation we shall often require an expression for the spectrum of radiation scattered from electrons with a range of energies. By analogy with eqn 2.16 this can be written as

$$J(\varepsilon)\,d\varepsilon = \int_E n(E)s(E,\varepsilon)\,dE\,d\varepsilon \qquad (2.31)$$

where $n(E)$ is the energy spectrum of the electrons and $s(\varepsilon, E)$ is given by eqn 2.23. In many problems we can replace the spectrum $s(E,\varepsilon)$ by a delta function at $\varepsilon = \langle\varepsilon\rangle$, where $\langle\varepsilon\rangle$ is given by eqn 2.24. Equation 2.31 then becomes

$$J(\varepsilon)\,d\varepsilon = \int_E n(E)P_T(E,\varepsilon)\,\delta(\varepsilon - \langle\varepsilon\rangle). \qquad (2.32)$$

When the energy spectrum of the electrons has the form of a power law

$$n(E)\,dE = n_0 E^{-m}\,dE$$

we can integrate eqn 2.32 and obtain

$$J(\varepsilon)\,d\varepsilon = J_0 \varepsilon^{-(m-1/2)}\,d\varepsilon \qquad (2.33)$$

where

$$J_0 = \frac{8\pi e^2 n_0 u}{9m^2 c^3 \varepsilon_0}\left(\frac{m^2 c^4}{2\varepsilon_0}\right)^{-(m-1/2)}$$

This spectrum has the same form as that derived in eqn 2.18 and again demonstrates the similarity between synchrotron radiation and the radiation produced by Thomson scattering.

It should be remembered that eqns 2.20 to 2.33 were derived under the assumption that $E\varepsilon_0 \ll m^2 c^4$. When $E\varepsilon_0 \gg m^2 c^4$ the interaction in the reference frame of the electron is Compton scattering and the expressions for the spectrum of the scattered radiation differ from those derived above.

In Compton scattering the photon always loses energy in the reference frame of the electron, and the energy of the scattered photon is approximately equal to $mc^2/2$ (Evans 1955). The angular distribution of the scattered radiation is given by the Klein–Nishima formula (Evans 1955); but, as in the case of Thomson scattering, we can get an approximate value for the mean energy of scattered photon in the observer's reference frame by considering just the extreme values of the scattering angle, θ^*. When $\theta^* = 0$, the energy of the scattered photon in the observer's reference frame is

$$\varepsilon = \varepsilon^*\sqrt{\left(\frac{1-\beta}{1+\beta}\right)} \sim \frac{mc^2}{4}\frac{mc^2}{E},$$

i.e.
$$\varepsilon \ll mc^2.$$

When $\theta^* = \pi$, the energy of the scattered photon in the observer's reference frame is

$$\varepsilon = \varepsilon^* \sqrt{\left(\frac{1+\beta}{1-\beta}\right)} \sim mc^2 \frac{E}{mc^2}$$

i.e. $$\varepsilon \sim E.$$

The mean energy of the scattered photon, and hence the energy lost by the electron, is given by

$$\varepsilon \sim \Delta E \sim \tfrac{1}{2}E. \tag{2.34}$$

The cross-section for Compton scattering is given (Evans 1955) by

$$\sigma_C = \frac{3mc^2\sigma_T}{8\varepsilon_0^*} \left\{ \ln\left(\frac{2\varepsilon_0^*}{mc^2}\right) + \frac{1}{2} \right\}.$$

Since

$$\varepsilon_0^* = \varepsilon_0 \frac{E}{mc^2}$$

we see that

$$\sigma_C = \frac{3m^2c^4\sigma_T}{8\varepsilon_0 E} \left\{ \ln\left(\frac{2\varepsilon_0 E}{m^2c^4}\right) + \frac{1}{2} \right\}. \tag{2.35}$$

The power lost by an electron will be given by

$$P_C(E, \varepsilon) \propto \Delta E \sigma_C. \tag{2.36}$$

From eqns 2.34 and 2.35 we see that, apart from the logarithmic term, $P_C(E, \varepsilon)$ is independent of the energy of the electron. When the electrons have a power law distribution of energies the spectrum of the scattered photons is therefore a power law with the same spectral index.

In some regions the density of relativistic electrons may be sufficiently high for a significant proportion of the photons to suffer two or more collisions before leaving the region. If we assume that a fraction, Δ, of the photons suffer at least one collision, then a fraction Δ^2 will suffer at least two collisions, and so on. We can write

$$\Delta \sim n\sigma R$$

where n is the density of relativistic electrons, σ is the scattering cross-section, and R is a typical dimension of the source. Provided $\varepsilon_0 E \ll m^2c^4$, the energy of the photon will be increased by a factor, g, where

$$g = \frac{4E^2}{3m^2c^4}. \tag{2.37}$$

After a number of collisions the energies of the photon will be such that $\varepsilon_0 E \gg m^2 c^4$ and we must then replace eqn 2.37 by

$$g \sim \frac{E}{2\varepsilon}.$$

Eventually, when $\varepsilon \sim E$, the photons will gain no further energy as a result of collisions.

The radiation emerging from such a region consists of several components, all of similar spectral shape, but occupying different regions of the spectrum. The frequencies in any component will be a factor g greater than those in the previous one, whilst the energy flux will be mutliplied by a factor $(g\Delta)$.

2.6. Bremsstrahlung

Bremsstrahlung is the radiation which is produced when an electron is accelerated in the electrostatic field of a nucleus or any other charged particle. The quantum mechanical treatment of this interaction has been given by Heitler (1944). Here we shall be content with a semiclassical treatment which demonstrates the essential features of the interaction.

Consider a relativistic electron, with total energy E and velocity βc which approaches to within a distance b of a nucleus with charge Ze. The problem can be treated most easily by considering the interaction in the reference frame of the electron because eqn 2.1 can then be used to estimate the power radiated. In the observer's reference frame the electrostatic field around the nucleus is isotropic and, at a distance b from the nucleus, the intensity is given by

$$\mathscr{E}_\perp \sim \frac{Ze}{b^2}$$

over a distance of the order of b. In the reference frame of the electron the electric field, $\mathscr{E}^*$, is given by

$$\mathscr{E}_\perp^* \sim \mathscr{E}_\perp \frac{E}{mc^2}.$$

Using eqn 2.1 we estimate the instantaneous power radiated by the electron, in either reference system, as

$$P_{\text{inst.}}(E, Z) = P_{\text{inst.}}^*(E, Z) \sim \frac{2Z^2 e^6}{3m^2 c^3 b^4}\left(\frac{E}{mc^2}\right)^2. \tag{2.38}$$

If the electron is moving through a medium in which there are N nuclei per cm^3 then the number of collisions which it experiences per second with impact parameter between b and $b + db$ is given by

$$n(b)\,db = 2\pi c N b\,db.$$

In each collision the power, given by eqn 2.38 is radiated for a period τ where

$$\tau \sim b/c.$$

Thus the average power radiated, as a result of all collisions, is

$$P_\mathrm{B}(E, Z) \sim \frac{4\pi N Z^2 e^6}{3m^2 c^3} \left(\frac{E}{mc^2}\right)^2 \int_{b_\mathrm{min}}^{\infty} \frac{\mathrm{d}b}{b^2}. \tag{2.39}$$

Integrating eqn 2.39 we find

$$P_\mathrm{B}(E, Z) \sim \frac{4\pi N Z^2 e^6}{3m^2 c^3} \left(\frac{E}{mc^2}\right)^2 \frac{1}{b_\mathrm{min}}. \tag{2.40}$$

Heitler (1944) has shown that the wave nature of the electron precludes collisions with impact parameters less than

$$b_\mathrm{min} \sim \frac{\hbar}{mc} \frac{E}{mc^2}.$$

Inserting this value of b_min into eqn 2.40 we obtain

$$P_\mathrm{B}(E, Z) \sim \frac{4\pi N Z^2 e^6}{3\hbar mc^2} \frac{E}{mc^2}. \tag{2.41}$$

The full quantum-mechanical treatment of bremsstrahlung gives (Heitler 1944)

$$P_\mathrm{B}(E, Z) = \frac{4 N Z^2 e^6}{\hbar mc^2} \frac{E}{mc^2} \ln\left(\frac{183}{Z^{1/3}}\right). \tag{2.42}$$

The radiation loss per unit length of path can be written as

$$\left(\frac{\mathrm{d}E}{\mathrm{d}x}\right)_\mathrm{B} = \frac{P_\mathrm{B}}{c} = \frac{E}{X_0},$$

where

$$X_0 = \left\{\frac{4 N Z^2 e^6}{\hbar m^2 c^5} \ln\left(\frac{183}{Z^{1/3}}\right)\right\}^{-1}. \tag{2.43}$$

X_0 is known as the radiation length.

In the reference frame of the electron, in addition to the electric field, $\mathscr{E}_\perp^*$, there is also a magnetic field, $\mathscr{H}_\phi^*$, whose intensity is given by

$$\mathscr{H}_\phi^* = \mathscr{E}_\perp \beta \frac{E}{mc^2} \sim \mathscr{E}_\perp^*.$$

Thus, to the electron, the electrostatic field of the nucleus appears as an electromagnetic pulse of duration

$$\tau^* \sim \tau \frac{mc^2}{E} \sim \frac{b}{c} \frac{mc^2}{E}.$$

The spectrum of the radiation produced can be derived by decomposing this pulse into its component frequencies and then considering the interaction as the scattering of these virtual photons by the electron. For convenience we shall consider the pulse to have a Gaussian shape with height, ZeE/b^2mc^2, and width, bmc/E. The Fourier transform of this pulse is another Gaussian with height, Ze/bc, and width, E/bmc. The interaction with the electron will be Thomson scattering for $v^* < v_c^*$, where

$$hv_c^* \sim mc^2,$$

and will be Compton scattering from $v^* > v_c^*$. Since the cross-section for Compton scattering is much smaller than that for Thomson scattering we shall ignore the interaction at frequencies greater than v_c^*. If the electron energy is large, the width of the Fourier transform of the pulse is much greater than v_c^* and we can then assume that the intensity, $j(v^*)$, of the virtual photons is independent of frequency up to v_c^*. The number spectrum of the photons which undergo scattering is therefore

$$n(v^*)\,dv^* = \frac{j(v^*)\,dv^*}{hv^*} \sim \frac{Ze}{hbc}\frac{dv^*}{v^*}.$$

The cross-section for Thomson scattering is independent of frequency, so the number spectrum of the scattered photons will have a similar shape with

$$n_s(v^*)\,dv^* \propto \frac{dv^*}{v^*}.$$

If the direction of the scattered photon makes an angle α to the direction of motion of the electron, when we transform back into the observer's reference frame we find

$$v = v^* \frac{1 + \beta \cos \alpha}{\sqrt{(1 - \beta^2)}}.$$

Therefore,

$$\frac{dv}{v} = \frac{dv^*}{v^*},$$

so that the spectrum of the radiation in the observer's reference frame is

$$n(v)\,dv \propto \frac{dv}{v}.$$

Obviously, the energy of the photon cannot be greater than the initial energy of the electron, so

$$v_{max} = E/h.$$

Thus the spectrum of the bremsstrahlung radiation is

$$j(E, v) = n(E, v)hv = j_0, v \leqslant E/h$$

$$= 0, v > E/h. \tag{2.44}$$

By comparing eqn 2.44 with eqn 2.42 we see that

$$j_0 \frac{E}{h} = P_B(E, Z),$$

i.e.
$$j_0 = \frac{8\pi N Z^2 e^6}{m^2 c^4} \ln\left(\frac{183}{Z^{1/3}}\right).$$

Consider now the radiation produced by an assembly of electrons with an energy spectrum, $n(E)$. The spectrum of the bremsstrahlung will be

$$J_B(v) \, dv = \int_{E=0}^{\infty} n(E)j(E, v) \, dE \, dv.$$

From eqn 2.44 we see that we need consider the integral only from $E = hv$ to $E = \infty$ and we can therefore write

$$J_B(v) \, dv = \int_{E=hv}^{\infty} n(E)j_0 \, dE \, dv. \tag{2.45}$$

2.7. Optical depth

The spectrum of the radiation which emerges from a source depends not only on the mechanism which is operating but also on the absorption characteristics of the source itself.

Consider a source in which the emission mechanism generates a spectrum $b(v)$ per unit volume. An element of the source which lies at a depth between x and $(x + dx)$ from the surface and which has unit cross-sectional area makes a contribution, $dJ(v)$, to the flux leaving the surface given by

$$dJ(v) = b(v) \exp\{-K(v)x\} \, dx, \tag{2.46}$$

where $K(v)$ is the attenuation coefficient of the radiation within the source.
Integrating eqn 2.46, we find

$$J(v) = \frac{b(v)}{K(v)} [1 - \exp\{-K(v)x_0\}], \tag{2.47}$$

where x_0 is the total depth of the source. The quantity $K(v)x_0$, which is usually written as $\tau(v)$, is known as the optical depth of the source.
We can distinguish two important cases. For $\tau(v) \ll 1$, we have

$$J(v) \sim b(v)x_0 \tag{2.48}$$

and the spectrum emitted by the source is identical to the spectrum generated by the radiation mechanism. For $\tau(v) \gg 1$, we have

$$J(v) \sim b(v)/K(v), \qquad (2.49)$$

and the spectrum at the surface is strongly modified by absorption processes within the source.

The concept of optical depth, and its value in calculations, can be illustrated by the derivation of the spectrum of thermal radiation from a hot plasma. The radiation is produced in collisions between the electrons and ions in the plasma and is therefore known as thermal bremsstrahlung. One way of treating this problem is to substitute the Boltzmann distribution of the energies of particles into eqn 2.45 but there is an alternate treatment which starts from the familiar black body spectrum of radiation.

Black body radiation is the thermal radiation emitted by a source which has a very large optical depth at all frequencies. The spectrum of black body radiation, which may be derived merely from the Bose–Einstein statistics of photons and without any consideration of the elementary processes involved, is given by

$$J_{\text{black body}}(v) = \frac{2hv^3}{c^2} \left\{ \frac{1}{\exp(hv/kT) - 1} \right\}.$$

Since thermal bremsstrahlung is the radiation emitted by a plasma with a very low optical depth, we see from eqns 2.48 and 2.49 that

$$J_{\text{thermal brems}}(v) \sim K(v) J_{\text{black body}}(v)$$

where $K(v)$ is the absorption coefficient of the photons in the plasma. The calculation of $K(v)$ requires a detailed consideration of the interaction of the photons with the particles of the plasma. Here, we shall merely quote the result,

$$K(T, v) = \frac{8\pi N^2 Z^2 e^6}{3hcm^2 v^3} g\sqrt{(m/6\pi kT)} \left\{ 1 - \exp\left(-\frac{hv}{kT} \right) \right\}.$$

Hence we see that

$$J(T, v)_{\text{thermal brems}} = \frac{16\pi N^2 Z^2 e^6}{3c^3 m^2} g\sqrt{(m/6\pi kT)} \exp\left(-\frac{hv}{kT} \right). \qquad (2.50)$$

The factor g, which is called the Gaunt factor, is only weakly dependent on frequency and temperature. The exponential term in eqn 2.50 ensures that the intensity falls rapidly at frequencies greater than kT/h and temperatures of the order of 10^9 K are required to produce significant fluxes of photons with energies greater than ~ 1 MeV.

References

Arnett, W. D. and Clayton, D. D. (1970). *Nature (London)* **227**, 780.

Bless, R. C. (1968). In *Stars and stellar systems* vol. VII, (ed. B. Middlehurst and L. H. Aller). University of Chicago Press, Chicago, p. 669.

Cohen, B. L. (1971). *Concepts of nuclear physics.* McGraw-Hill, New York.

Evans, R. D. (1955). *The atomic nucleus.* McGraw-Hill, New York.

Ginzburg, V. L. (1969). *Elementary processes for cosmic ray astrophysics.* Gordon and Breach, New York.

Hayward, R. W. (1967). In *Handbook of Physics* (ed. E. U. Condon and H. Odishaw). McGraw-Hill, New York.

Heitler, W. (1944). *The quantum theory of radiation,* 1st edn. Oxford University Press, London.

Jackson, D. J. (1962). *Classical electrodynamics.* John Wiley and Sons, Inc., New York.

Lederer, C. M., Hollander, J. M., and Perlman, I. (1968). *Table of isotopes.* John Wiley, New York.

Ramaty, R., Kozbovsky, B., and Lingenfelter, R. E. (1979). *Astrophys. J. Suppl. Ser.* **40**, 487.

Stecker, F. W. (1971). *Cosmic gamma rays.* Mono Book Corporation, Baltimore, MD.

Tucker, W. H. (1975). *Radiation processes in astrophysics.* MIT Press, Cambridge, MA.

3

THE PROBLEM OF BACKGROUND
RADIATION IN GAMMA RAY ASTRONOMY

3.1. Introduction

Astronomers who work in the optical or radio regions of the electromagnetic spectrum are privileged insofar as they can make their measurements from the surface of the earth with little hindrance from the earth's atmosphere. Those who work at other frequencies are faced with the problem of atmospheric attenuation and must arrange for their detectors to be carried to the top of the atmosphere. Moreover, at some frequencies the atmosphere provides a further obstacle by acting as a source of background radiation against which the weak fluxes from astronomical sources must be detected; for example scattered sunlight precludes most optical observations during daytime and thermal emission from the atmosphere poses a problem for infra-red astronomers. In the gamma ray region of the spectrum a large background flux is produced by the interaction of the cosmic radiation with the atmosphere. These problems of atmospheric attenuation and atmospheric background radiation are the basic ones in experimental gamma ray astronomy and they must be overcome before significant astronomical observations can be made.

3.2. Atmospheric attenuation of gamma rays

At photon energies below ~ 60 keV, where the photoelectric effect predominates, and at energies above ~ 10 MeV, where pair production predominates, the definition of the mass attenuation coefficient, $\sigma(v)$, is unambiguous and is given by the equation

$$I(v) = I_0(v) \exp\{-\sigma(v)x\rho\},$$

where $I_0(v)$ is the initial intensity of photons and $I(v)$ is the intensity after travelling a distance, x, through air of density, ρ. At energies between ~ 60 keV and ~ 10 MeV the majority of photons interact with air through Compton scattering. In this process the scattered photon emerges from the interaction, but with its energy reduced; in some situations it is the attentuation of the energy carried by the photons which is important but in astronomy we are usually interested in the attenuation of the flux of unscattered photons. The photons which are scattered and continue with reduced energy may subsequently be detected and a correction must be applied for these.

Figure 3.1 shows the mass attenuation coefficients in air for photons with energies between 10 keV and 100 MeV. The coefficient, τ, refers to the photoelectric effect and the coefficient, κ, to pair-production. Two coefficients, σ_a and σ_s, are shown for Compton scattering and these are defined in the following way. If a beam of photons of energy ε undergoes Compton scattering then the photons which are scattered will suffer, on average, an energy loss $\Delta\varepsilon$. If σ_c is the total coefficient for Compton scattering then σ_a, σ_s are defined by

$$\sigma_a = \sigma_c \frac{\Delta\varepsilon}{\varepsilon}$$

and

$$\sigma_s = \sigma_c \left(1 - \frac{\Delta\varepsilon}{\varepsilon}\right).$$

So, σ_a refers to the energy absorbed and σ_s to the energy scattered. Obviously,

$$\sigma_c = \sigma_a + \sigma_s.$$

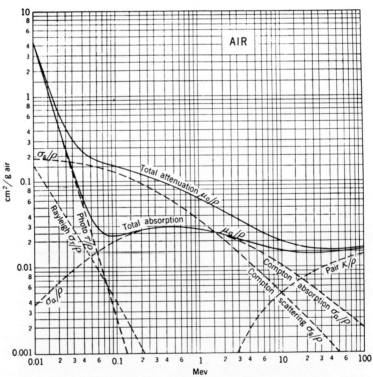

FIG. 3.1. Attenuation coefficients for photons in air. (From Evans 1955, p. 713.)
© (1955) McGraw-Hill Book Company.

The coefficients μ_a, μ_0 which are also shown in Fig. 3.1 are defined by

$$\mu_a = \tau + \sigma_a + \kappa$$

and

$$\mu_0 = \tau + \sigma_a + \sigma_s + \kappa.$$

It is clear that to avoid excessive attenuation at energies above 0.5 MeV astronomical measurements must be made within $\sim 10\ \mathrm{g\ cm^{-2}}$ at the top of the atmosphere. In Fig. 3.2 the atmospheric depth, in $\mathrm{g\ cm^{-2}}$, is plotted against altitude, in km; from Fig. 3.1 and Fig. 3.2 we see that measurements must be made at altitudes above 100 km. This implies that observations can be made from balloons, rockets, or satellites, but not from conventional aircraft. In Chapter 6 we shall see that the interaction of a very high energy photon gives rise to an electromagnetic cascade which penetrates large distances into the atmosphere and at photon energies above $\sim 10^{11}$ eV, these secondary effects allow the detection of primary photons at ground level.

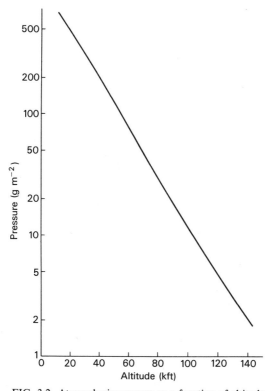

FIG. 3.2. Atmospheric pressure as a function of altitude.

3.3. Background effects produced by the cosmic radiation

3.3.1. *Introduction*

The pioneering experiments in gamma ray astronomy were made with detectors carried up into the stratosphere by large balloons. These experiments showed that the greatest experimental obstacle to be overcome in this field of astronomy is the large flux of gamma rays, at all energies, which are produced as a result of collisions of cosmic ray particles in the atmosphere (Fig. 3.3). Considerable effort was made to understand the nature of this background and the mechanisms by which it is produced. Later experiments used detectors carried on artificial earth satellites but, although these

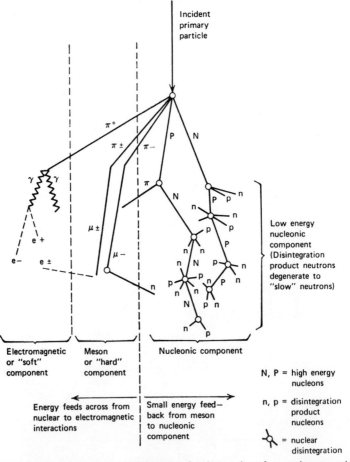

FIG. 3.3. Schematic representation of the results of an interaction of a cosmic ray particle with the atmosphere. (From Simpson, Fonger, and Treiman 1953, p. 936.)

detectors were outside the atmosphere, they were not free from background radiation. For satellites in orbits near the earth there was still a large upward-moving component of the atmospheric radiation and for satellites in deep space there was background from the spacecraft itself, generated again from the interactions of cosmic ray particles. In this discussion of the background radiation we shall concentrate on the production of radiation in the earth's atmosphere; similar processes will take place in the body of a spacecraft although the detailed nature of the radiation will depend on the precise form of the spacecraft.

3.3.2. *The primary cosmic radiation*

The primary cosmic radiation consists principally of high energy protons. The energy spectrum of the protons, well away from the earth's magnetic field, has the form shown in Fig. 3.4. At energies greater than ~ 10 GeV the spectrum has the form of a power law,

$$n(E)\,\mathrm{d}E = KE^{-m}\,\mathrm{d}E$$

with $$m = \sim 2.6.$$

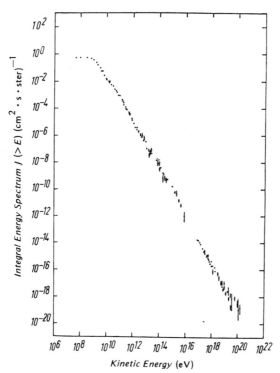

FIG. 3.4. Integral energy spectrum of the cosmic ray protons. (From Unsold 1969, p. 349.)

The spectrum observed within the solar system below ~10 GeV falls below the power law distribution, but this is at least partly due to modulation by inplanetary magnetic fields. The magnitude of this modulation varies through the eleven-year cycle of solar activity as shown in Fig. 3.5.

At the top of the earth's atmosphere the spectrum is also modified by the

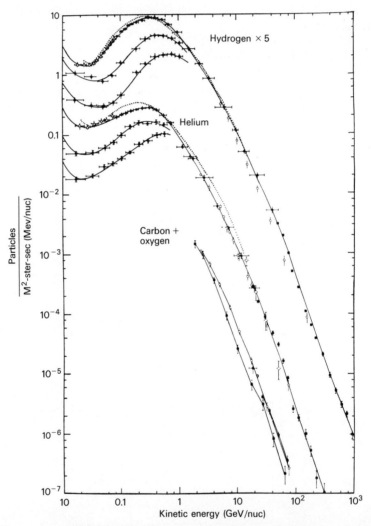

FIG. 3.5. Energy spectra of primary cosmic ray particles. At energies below ~1 GeV per nucleon the intensities are modulated by interplanetary magnetic fields and spectra are shown for three different phases in the 11-year cycle of solar activity; the intensities are greatest when solar activity is at a minimum. (From Webber and Lezniak 1974, p. 363.)

earth's magnetic field. This introduces a low energy cut-off in the spectrum, below which protons are unable to reach the earth's atmosphere. For a simple dipole field the momentum μ_c, corresponding to the cut-off energy, is given by

$$\mu_c = 15 \cos^4 \lambda_G,$$

where μ_c is measured in GeV/c. More accurate values of the cut-off momenta at different locations on the earth, derived from the real magnetic field, have been given by Quenby and Webber (1959). The geomagnetic cut-off causes the intensity of the primary radiation to decrease with decreasing magnetic latitude.

The primary cosmic radiation also contains energetic nuclei of helium and the heavier elements. Helium nuclei make up only some 12 per cent of the particles, but since each nucleus contains four nucleons, helium nuclei do represent some 30 per cent of the nucleons in the primary radiation and their effects are not negligible. The energy spectrum of the helium nuclei, when expressed in energy per nucleon, has a form similar to that of the protons, as can be seen in Fig. 3.5. The nuclei of elements heavier than helium are not important as a source of background radiation.

The intensity of the secondary radiation produced in the atmosphere varies with geomagnetic latitude, but the variation is less than that of the primary radiation because the low energy primary particles, which can reach the atmosphere only at higher latitudes, are less effective in producing secondary radiation. Staib, Frye, and Zych (1974) have reported that the atmospheric flux of high energy gamma rays, with energies greater than ~ 50 MeV, at the geomagnetic equator is approximately half that at high latitudes; Kasturirangan, Rao, and Bhavsar (1972) claim that the flux of low energy gamma rays decreases by a factor of about nine between the same latitudes.

3.3.3. *High energy secondary protons, neutrons, and pions in the atmosphere*

When a cosmic ray nucleus collides with a nucleus in the atmosphere some of the nucleons from either nucleus may emerge from the interaction with energies of several hundred MeV or greater. These nucleons, together with the charged pions created in the interaction, form a secondary component which is capable of inducing further high energy nuclear disintegrations. Consequently the number of high energy interactions per unit mass of the atmosphere increases with atmospheric depth; but, in successive collisions, the energies of the secondary particles steadily decrease and eventually they are incapable of inducing further disintegrations. The intensities of all components of the background radiation follow the intensity of the high energy particles; each component has a maximum at some depth in the atmosphere, often referred to as the Pfotzer maximum, and below this level the intensity decreases exponentially with depth with an attenuation length of

$\sim 140 \text{ g cm}^{-2}$, this being the attenuation length of the flux of high energy particles deep in the atmosphere.

3.3.4. *Background radiation from the decay of secondary pions*

We have seen in Chapter 2 that the decay of pions is an important source of high energy gamma rays. In the earth's atmosphere both charged and neutral pions are created in the nuclear collisions between cosmic ray particles and the atmosphere. A charged pion decays first to a muon and then to an electron,

$$\pi^\pm \to \mu^\pm + \nu + \bar{\nu},$$
$$\mu^\pm \to e^\pm + \nu + \bar{\nu}.$$

A neutral pion decays into two gamma rays,

$$\pi^0 \to h\nu + h\nu.$$

The gamma rays subsequently interact with the atmosphere through pair production, giving relativistic electrons.

$$h\nu \to e^+ + e^-.$$

These electrons, and those from the decay of the charged pions, then produce further gamma rays through bremsstrahlung,

$$e^\pm + (N, Z) \to e^\pm + (N, Z) + h\nu.$$

In this way the energy of the pions is converted into a cascade of electrons and gamma rays. As the cascade moves down through the atmosphere the average energies of the electrons and gamma rays decrease. Electrons with energies below 80 MeV lose their energy by ionization rather than by bremsstrahlung, and gamma rays with energies below 10 MeV tend to interact through Compton scattering, rather than pair production. When these energies are reached the cascade ceases to grow. The low energy electrons and positrons quickly lose their energy by ionization and the positrons eventually annihilate at rest, producing a line spectrum of gamma rays at 0.511 MeV; the low energy gamma rays are much more penetrating and form a major component of the gamma ray background in the atmosphere.

At energies above ~ 1 GeV the major contribution to the background gamma ray flux comes directly from the decay of neutral pions. In Chapter 2 we saw that, if the primary protons have an energy spectrum of the form

$$n(E) \propto E^{-m}$$

then the gamma rays resulting from the decay of the secondary pions will have an energy spectrum

$$n(\varepsilon) \propto \varepsilon^{-p}$$

where, at high energies, $p \sim \frac{4}{3}(m - \frac{1}{2})$. Since m is equal to 2.6 we calculate p should be equal to 2.8; the measured value is 2.6 (Anand, Daniel, and Stephens 1973). A high energy gamma ray produced directly from the decay of a neutral pion travels in a direction which is closely correlated with that of the primary cosmic ray particle which produced the initial nuclear disintegration. For this reason the flux of high energy gamma rays, near the top of the atmosphere, is far from isotropic. There are few upward-moving gamma rays because there are no upward-moving primary particles and there are few vertically downward-moving gamma rays because there is little atmosphere directly overhead in which nuclear disintegrations can occur; the gamma ray flux is therefore a maximum at the horizon (Fig. 3.6).

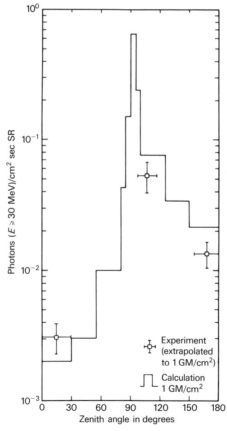

FIG. 3.6. Zenith angle dependence of the intensity of background gamma rays with energies greater than ~ 30 MeV at an atmospheric depth of 1 g cm^{-2}. (From Thompson 1974, p. 1318.)

Below ~ 1 GeV bremsstrahlung from electrons in the electromagnetic cascade becomes significant. The spectrum from pion decay reaches a maximum at 67.5 MeV and the decreasing contribution from this component at lower energies produces a flattening of the spectrum.

At energies below ~ 1 MeV the photon energy spectrum is determined by the rate of energy loss of the photons by Compton scattering (Griffiths 1979). The number of photons, $n(\varepsilon, t)$, with energies between ε and $\varepsilon + \mathrm{d}\varepsilon$ will be given by the solution of the Boltzmann equation,

$$\frac{\mathrm{d}}{\mathrm{d}t}\{n(\varepsilon, t)\} = \frac{\partial}{\partial \varepsilon}\left\{n(\varepsilon, t)\frac{\mathrm{d}\varepsilon}{\mathrm{d}t}\right\} + Q(\varepsilon, t).$$

The first term on the r.h.s. takes account of photons entering and leaving the energy interval because of energy losses by Compton scattering. The second term on the r.h.s. represents the direct gain or loss of photons due to production or absorption mechanisms; at low energies the most important of these is the absorption of photons through the photoelectric effect. For this process we may write

$$Q(\varepsilon, t) = -\tau(\varepsilon)cn(\varepsilon, t),$$

where $\tau(\varepsilon)$ is the linear absorption coefficient for the photoelectric effect. In equilibrium we may write

$$\frac{\mathrm{d}}{\mathrm{d}t}\{n(\varepsilon, t)\} = 0$$

i.e.

$$n(\varepsilon)\frac{\partial}{\partial \varepsilon}\left\{\frac{\mathrm{d}\varepsilon}{\mathrm{d}t}\right\} + \frac{\partial n(\varepsilon)}{\partial \varepsilon}\left\{\frac{\mathrm{d}\varepsilon}{\mathrm{d}t}\right\} - \tau(\varepsilon)cn(\varepsilon) = 0.$$

There will be a maximum in the equilibrium spectrum, at low energies, where

$$\frac{\partial n(\varepsilon)}{\partial \varepsilon} = 0$$

i.e.

$$n(\varepsilon)\frac{\partial}{\partial \varepsilon}\left\{\frac{\mathrm{d}\varepsilon}{\mathrm{d}t}\right\} - \tau(\varepsilon)cn(\varepsilon) = 0$$

or

$$\frac{\partial}{\partial \varepsilon}\left\{\frac{\mathrm{d}\varepsilon}{\mathrm{d}t}\right\} = \tau(\varepsilon)c. \tag{3.1}$$

$\left\{\dfrac{\mathrm{d}\varepsilon}{\mathrm{d}t}\right\}$ is related to the Compton absorption coefficient, $\sigma_{\mathrm{a}}(\varepsilon)$, in Fig. 3.1 by

$$\left\{\frac{\mathrm{d}\varepsilon}{\mathrm{d}t}\right\} = \sigma_{\mathrm{a}}(\varepsilon)\varepsilon c.$$

Thus,

$$\frac{\partial}{\partial \varepsilon} \left\{ \frac{d\varepsilon}{dt} \right\} = c\sigma_a(\varepsilon) + c\varepsilon \frac{\partial \sigma_a(\varepsilon)}{\partial \varepsilon}.$$

Now $\sigma_a(\varepsilon)$ is a slowly varying function of ε and to a first approximation we may neglect the second term on the r.h.s. The maximum in the energy spectrum of the photons, defined by eqn 3.1, therefore occurs where

$$\tau(\varepsilon) \sim \sigma_a(\varepsilon).$$

We see from Fig. 3.1 that the maximum intensity in air occurs at an energy of ~ 60 keV. It is only at photon energies below 60 keV that the astronomer can expect to gain some respite from the problems of atmospheric background radiation.

Many attempts have been made to shield detectors from the large flux of low energy gamma rays using collimators of lead or some other material with high atomic number and high photoelectric cross-sections. The processes which we have discussed above then take place in the collimator immediately around the detector as well as in the atmosphere outside. The spectrum of the secondary radiation will be very little different at high energies, but at low energies the position of the maximum in the spectrum is changed because of the higher photoelectric absorption in the material of the collimator.

Figure 3.7 shows the mass attenuation coefficients for lead. We see that the background spectrum inside a collimator constructed from lead would have a maximum at ~ 900 keV and at energies below this the intensity of the background radiation would be less than in the air outside; it is difficult, by passing shielding alone, to reduce the intensity at energies above ~ 900 keV and, indeed, significant reductions do not occur above ~ 300 keV.

3.3.5. *The inelastic scattering and capture of low energy neutrons*

An oxygen or nitrogen nucleus which has been struck by a cosmic ray particle is often left with a very large excitation energy which it subsequently releases in the form of protons or neutrons with energies up to ~ 10 MeV. These nucleons are often referred to as 'evaporation particles' because their release resembles the evaporation of molecules from a hot liquid drop. The protons are quickly brought to rest by ionization but the neutrons are more penetrating and lose their energy only by inelastic nuclear collisions of the type

$$n + (N, Z) \rightarrow (N, Z)^* + n'.$$

The target nucleus, which is left in an excited state, may return to the ground state with the emission of a gamma ray. Eventually the neutron will be captured, again giving a nucleus in an excited state,

$$n + (N, Z) \rightarrow (N + 1, Z)^*.$$

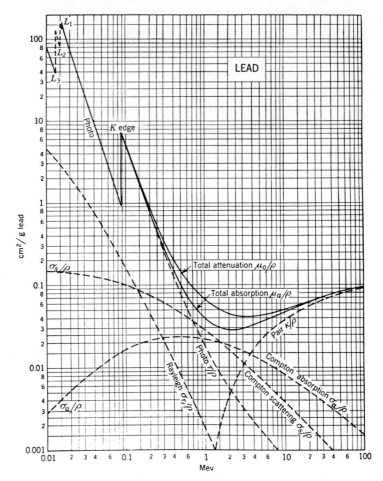

FIG. 3.7. Attenuation coefficients for photons in lead. (From Evans 1955, p. 716.) © (1955) McGraw-Hill Book Company.

The low energy neutrons in the atmosphere have an energy spectrum which, like that of the low energy gamma rays, is determined by the rate at which they lose energy in collisions. Figure 3.8 shows the spectra calculated by Armstrong, Chandler, and Barish (1973) for different depths in the atmosphere. The gamma rays emitted from the excited nuclei have line spectra but, before reaching the detector, they may suffer many Compton collisions and the background contribution from these processes will be a continuum with relatively weak lines superimposed on it. The details of this spectrum depends on the nature of the material immediately around the detector.

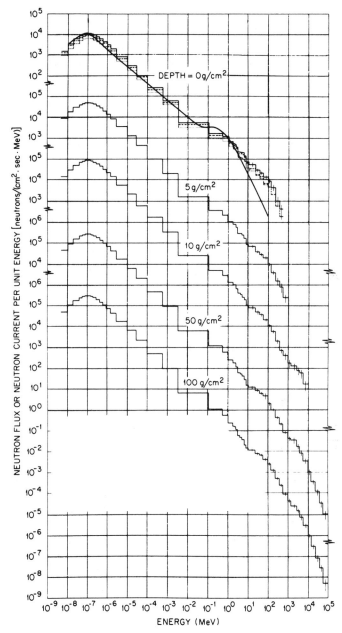

FIG. 3.8. Energy spectra of atmospheric neutrons at various atmospheric depths. The spectra shown are for minimum solar activity and at a geomagnetic latitude $\lambda = 42°$N. (From Armstrong *et al.* 1973, p. 2719.)

3.3.6. *Induced radioactivity*

The level of natural radioactivity in the materials used in a gamma ray telescope is usually too low for this process to make a significant contribution to the background counting rate in the detector. But induced radioactivity, due to nuclei produced in interactions of cosmic ray particles may be a serious problem; this is especially true for detectors in satellites, which are subjected to long exposures to the cosmic radiation and which may pass repeatedly through the earth's radiation belts.

The effects of many forms of background radiation which are produced in the interactions of cosmic ray particles can be eliminated or reduced by incorporating further detectors in the shielding of the main detector; such a configuration is often referred to as an active shield and will be discussed in the next chapter. Any event in the main detector which is associated, in time, with the passage of a charged particle through the shield is rejected. However, this technique is ineffective against induced radioactivity because there may be a long delay, of up to several days, between the event which produces the radioactive nucleus and its subsequent decay.

The spectrum of the background which is produced from this process depends on the materials used in the construction of the telescope, but the energy release is seldom more than a few MeV. It should be noted that when the radioactive decay occurs inside a detector such as a scintillation counter it may be counted as a background event even though no gamma ray is emitted; for example, the emission of an electron in a β-decay is indistinguishable in most simple detectors from the absorption of a gamma ray.

Dyer and Morfill (1971) have calculated the radioactivity produced in a cesium iodide crystal which has been exposed to the primary cosmic radiation and to the particles in the radiation belts. They used an empirical formula developed by Rudstam (1966) to predict the spallation products from the nuclear disintegrations, and they concluded that the most copiously produced radioactive species would be the neutron-deficient isotopes of iodine and cesium. A further source of radioactive nuclei is the absorption of slow neutrons by ^{128}I to form ^{129}I. The net result of the subsequent decay of these radioactive nuclides is a spectrum of energy losses which is a continuum with a strong peak at ~ 200 keV due to the decay of ^{123}I by electron capture. Fishman (1973) measured experimentally the radioactivity produced in a sodium iodide crystal which was irradiated with 600 MeV protons from an accelerator. He found the energy loss spectrum could be represented by

$$n(\varepsilon)\,d\varepsilon \sim \exp(\varepsilon/\varepsilon_0)\,d\varepsilon,$$

where

$$\varepsilon_0 \sim 1\,\text{MeV}.$$

3.3.7. *The background radiation at balloon altitudes at a geomagnetic latitude,* $\lambda \sim 40°$

In the previous sections we have seen that the background radiation has several components, each produced by a different mechanism; moreover the intenstiy varies with geomagnetic latitude and depends on whether the detector is at balloon altitudes or in space. To illustrate the practical problems which this radiation poses for the gamma ray astronomer we shall consider here the background in a particular situation, namely at an atmospheric depth of ~ 5 g cm^2 and at a geomagnetic latitude, $\lambda \sim 40°$. It will be convenient to discuss the background radiation at high and low energies separately.

(a) *Background radiation at energies greater than ~ 30 MeV.* The detectors for this region of the spectrum, which are usually spark chambers, can reliably distinguish gamma rays from other forms of background and they are also capable of measuring directly the direction of motion of a gamma ray. Figure 3.9 shows the results of a number of measurements, at $\lambda \sim 40°$, of the intensity around the vertical direction at atmospheric depths of a few g cm^{-2}. Since the intensity of the background gamma ray flux at these energies is directly proportional to the mass of overlying air the intensities have been quoted per unit mass of air.

(b) *Background radiation at energies less than ~ 30 MeV.* At these energies it is difficult, without the use of a collimator, to measure the direction of motion of a photon. Unfortunately, the addition of a collimator inevitably changes the background radiation at the detector in a way which is difficult to calculate. We shall therefore present the background recorded by an unshielded detector and this will obviously correspond to radiation received over the full solid angle.

Figure 3.10 shows the spectrum recorded by Peterson *et al.* (1973), again at $\lambda \sim 40°$, using a 3 in diameter $\times$ 3 in high sodium iodide scintillation crystal at an atmospheric depth of ~ 4 g cm^2; the counting rate at these energies is not strongly dependent on atmospheric depth. No attempt has been made to convert the counting rates to photon fluxes since, as we have seen in Section 3.3.6, some of the events may not be caused by gamma rays. The only spectral line which is evident is that at 0.511 MeV arising from the annihilation of positrons created in electromagnetic cascades in the atmosphere.

3.4. The limiting sensitivity of a gamma ray telescope

An astronomical source is detected as an increase in flux, sometimes very small, over the background flux expected from a particular direction in the

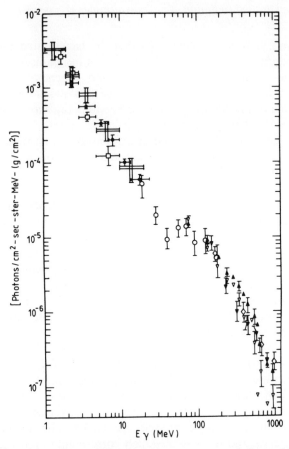

FIG. 3.9. Background spectrum of high energy gamma rays at balloon altitudes and at a geomagnetic latitude $\lambda = 40°$. (From Schonfelder, Graml and Penningsfeld 1980, p. 360.) © (1980) The American Astronomical Society.

sky. The ultimate sensitivity of any telescope is set by the statistical fluctuations in the number of photons intercepted. In many regions of the electromagnetic spectrum the number of photons detected in a typical measurement is so large that this limit to the sensitivity is seldom reached and other factors are important. But, at gamma ray frequencies the energy carried by a single photon is very large and the number of photons in a beam of a given energy flux is correspondingly small. The statistical fluctuations in the total number of photons detected, both in the signal and in the background radiation, are thus the overriding factor in the sensitivity of a gamma ray telescope.

Consider a set of measurements on a source which produces at the earth a

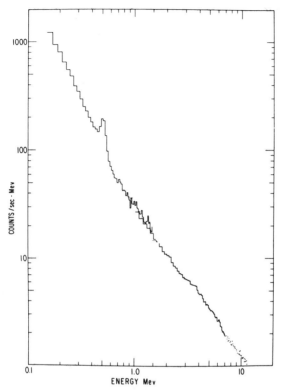

FIG. 3.10. Background counting rate in a sodium iodide scintillation crystal 3 in. diameter × 3 in. thick at balloon altitudes and at a geomagnetic latitude $\lambda = 40°$. (From Peterson *et al.* 1973, p. 7947.)

flux of photons, $S(\varepsilon)$, with energies between ε and $\varepsilon + d\varepsilon$. We shall suppose the spectrum from the source to be a continuum and the detector to accept photons over an energy interval $\Delta\varepsilon$, centred on ε. Let the sensitive area of the detector by A, its efficiency be $\eta(\varepsilon)$ and the background counting rate per unit area be $b(\varepsilon)$.

A typical measurement will consist of a period T_s spent recording photons from the direction of the source and a time T_b spent recording photons from neighbouring directions in the sky. The number of events recorded when the telescope is pointing towards the source is given by

$$N_s = \{S(\varepsilon)\eta(\varepsilon) + b(\varepsilon)\}A\,\Delta\varepsilon T_s.$$

If the number of photons, N_s, is large its standard deviation due to Poissonian fluctuations is given by

$$\Delta N_s = \pm\sqrt{[\{S(\varepsilon)\eta(\varepsilon) + b(\varepsilon)\}A\,\Delta\varepsilon T_s]}.$$

The counting rate, n_s, recorded from the direction of the source can therefore be written as

$$n_s = \{S(\varepsilon)\eta(\varepsilon) + b(\varepsilon)\}A\ \Delta\varepsilon \pm \sqrt{\left(\frac{\{S(\varepsilon)\eta(\varepsilon) + b(\varepsilon)\}A\ \Delta\varepsilon}{T_s}\right)}.$$

Similarly, when the telescope is pointing towards a neighbouring direction in the sky the counting rate is

$$n_b = b(\varepsilon)A\ \Delta\varepsilon \pm \sqrt{\left(\frac{b(\varepsilon)A\ \Delta\varepsilon}{T_b}\right)}.$$

The signal from the source is seen as the difference between these two counting rates which can be written as

$$n_s - n_b = S(\varepsilon)\eta(\varepsilon)A\ \Delta\varepsilon \pm \sqrt{\left[\left\{\frac{S(\varepsilon)\eta(\varepsilon) + b(\varepsilon)}{T_s} + \frac{b(\varepsilon)}{T_b}\right\}A\ \Delta\varepsilon\right]}.$$

The difference between the rates is significantly greater than zero, and the source is detected, only if

$$S(\varepsilon)\eta(\varepsilon)A\ \Delta\varepsilon \geqslant \beta\sqrt{\left[\left\{\frac{S(\varepsilon)\eta(\varepsilon) + b(\varepsilon)}{T_s} + \frac{b(\varepsilon)}{T_b}\right\}A\ \Delta\varepsilon\right]}, \qquad (3.2)$$

where β is a small number which expresses the degree of confidence required in the measurement. Typically, β is equal to 3.

In practice the fluxes of photons recorded from sources are usually very much smaller than the background counting rates and in this approximation eqn 3.2 becomes

$$S(\varepsilon) \geqslant \frac{\beta}{\eta(\varepsilon)}\sqrt{\left\{\frac{(1/T_s + 1/T_b)b(\varepsilon)}{A\ \Delta\varepsilon}\right\}}. \qquad (3.3)$$

If T is the total observing time for the measurement,

$$T = T_s + T_b$$

and we may rewrite eqn 3.3 as

$$S(\varepsilon) \geqslant \frac{\beta}{\eta(\varepsilon)}\sqrt{\left\{\left(\frac{1}{T_s} - \frac{1}{T - T_s}\right)\frac{b(\varepsilon)}{A\ \Delta\varepsilon}\right\}}.$$

In many experiments the total observing time, T, is fixed and the greatest sensitivity is then achieved when

$$T_s = T/2;$$

in these circumstances

$$S(\varepsilon) \geqslant \frac{\beta}{\eta(\varepsilon)}\sqrt{\left\{\frac{2b(\varepsilon)}{A\ \Delta\varepsilon T_s}\right\}}.$$

References

Anand, K. C., Daniel, R. R., and Stephens, S. A. (1973). *Pramana* **1**, 2.

Armstrong, T. W., Chandler, K. C., and Barish, J. (1973). *J. geophys. Res.* **78**, 2715.

Dyer, C. S. and Morfill, G. E. (1971). *Astrophys. space Sci.* **14**, 243.

Evans, R. D. (1955). *The atomic nucleus.* McGraw-Hill, New York.

Fishman, G. J. (1973). In *Gamma ray astrophysics, NASA Spec. Publ. SP339*, p. 61.

Griffiths, H. (1979). Ph.D. thesis. University of Bristol, Bristol.

Kasturirangan, K., Rao, U. R., and Bhavsar, P. D. (1972). *Planet. Space Sci.* **20**, 1961.

Peterson, L. E., Swartz, D. A., and Ling, J. C. (1973). *J. geophys. Res.* **78**, 7942.

Quenby, J. J. and Webber, W. R. (1959). *Philos. Mag.* **4**, 90.

Rudstam, G. (1966). *Z. Naturforsch.* **21a**, 1027.

Schonfelder, V., Graml, F., and Penningsfeld, F.-P. (1980). *Astrophys. J.* **240**, 350.

Simpson, J. A., Fonger, W., and Treiman, S. B. (1953). *Phys. Rev.* **90**, 934.

Staib, J. A., Frye, G. M., and Zych, A. D. (1974). *J. geophys. Res.* **79**, 929.

Thompson, D. J. (1974). *J. geophys. Res.* **79**, 1309.

Unsold, A. (1969). *The new cosmos.* Longmans Springer-Verlag, New York.

Webber, W. R. and Lezniak, J. A. (1974). *Astrophys. space Sci.* **30**, 361.

4

LOW ENERGY GAMMA RAY TELESCOPES

4.1. Introduction

All gamma ray detectors rely on a measurement of the electron which is ejected when a gamma ray interacts with matter. The design of a gamma ray telescope to operate in a particular region of the spectrum is therefore largely dictated by the mechanism through which the gamma rays interact. In this chapter we shall consider the design of telescopes for photons with energies between 0.5 MeV and ~ 3 MeV and we see from Fig. 4.1 that throughout this energy interval Compton scattering is the most important interaction mechanism in all materials.

A telescope is required, not only to detect photons, but also to measure their directions of motion. The direction at which the electron is ejected in Compton scattering is only weakly correlated with the direction of motion of the gamma ray; moreover the low energy electron suffers severe multiple scattering in the detector and its initial direction of motion is soon lost. Consequently very little success has been achieved in the construction of low energy gamma ray detectors with inherent directional sensitivity. Instead, telescopes usually consist of a detector, which merely measures the energy of the gamma ray, placed behind some form of collimator. Even the measurement of the gamma ray energy presents some difficulties because the gamma ray loses only a fraction of its energy in a single Compton scattering and

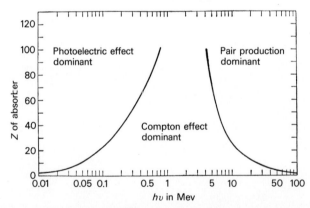

FIG. 4.1. Relative importance of the different forms of gamma ray interaction as a function of photon energy and the atomic number of the material. (From Evans 1955, p. 712.) © (1955) McGraw-Hill Book Company.

many scatterings, terminated by a photoelectric absorption, are required if the total energy of the gamma ray is to be absorbed in the detector. The cross-section for Compton scattering is not large and a detector needs to be massive if it is to be efficient. For this reason the detector usually takes the form of a scintillation counter.

4.2. Scintillation counters

The basic components of a scintillation counter are a scintillator, which converts into light a fraction of the energy lost by a charged particle as ionization, and a photoelectric device which converts the light into an electrical signal. The materials which are readily available as scintillators are either inorganic crystals, such as sodium iodide and caesium iodide, or plastics loaded with suitable scintillating organic compounds. The inorganic crystals are usually favoured in laboratory experiments which involve gamma ray spectroscopy, because iodine and caesium, with high atomic numbers, have large cross-sections for photoelectric absorption. But the cost of the pure materials for these scintillators is high and the growing of large single crystals is difficult. By contrast, plastic scintillator is much less expensive and can readily be manufactured and machined to meet the needs of particular experiments.

The scintillator in many low energy gamma ray detectors is in the form of a cylinder. For small detectors a photomultiplier is optically coupled to one of the flat faces of the cylinder and the other faces are covered with a highly reflecting material such as aluminium foil or white paint. In such an arrangement the efficiency of light collection by the photomultiplier is very high, often exceeding ~90 per cent. When the scintillator is very large only a small fraction of it may be covered by photomultiplier and it is then much more difficult to achieve efficient and uniform light collection. Several methods of collecting the light from a large scintillator are shown in Fig. 4.2;

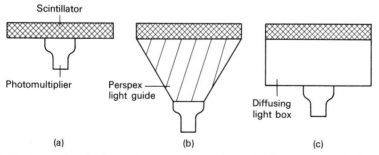

FIG. 4.2. Different methods of optically coupling a photomultiplier to a large scintillator: (a) direct coupling; (b) coupling through a Perspex light guide; (c) coupling through a diffusing light box.

the technique which is chosen will depend on the importance placed on particular characteristics of the detector. For example in method (a) where the photomultiplier is mounted directly on the scintillator the design is compact but the efficiency of light collection varies considerably over the area of the scintillator; in method (b) where a light guide is used to concentrate the light onto the photomultiplier the light collection is uniformly high, but the mass of the light guide is considerable; and in method (c), where a box with diffusing sides is used to retain the light, the efficiency of light collection is low but it is uniform and the mass of the system is low.

The photocathodes of photomultipliers have quantum efficiencies ranging from ~ 10 to ~ 20 per cent. The statistical fluctuation in the number of photoelectrons liberated from the photocathode for a given energy deposited in the scintillator is one of the factors governing the energy resolution of a scintillation counter; the other important factors are the non-uniformity of light collection from the scintillator and the variation in quantum efficiency over the area of the photocathode. With good efficiency of light collection a 1 MeV energy loss in a sodium iodide crystal will result in the emission of ~ 4000 photoelectrons from the photocathode and the statistical fluctuation in this number will contribute less than 2 per cent to the spread of the output signals from the photomultiplier. In the case of a large scintillator the variation of light collection is usually the factor which determines the energy resolution of the detector.

4.3. Solid state detectors

When energy resolution, rather than large area, is of prime importance solid state detectors offer a considerable advantage over scintillation counters. The most convenient solid state detectors for gamma ray spectroscopy are those constructed from lithium drifted silicon or germanium because they contain relatively large volumes of active material. Both types of detectors must be operated at low temperatures if the maximum energy resolution is required; germanium detectors suffer the disadvantage that they must also be stored at low temperatures to prevent permanent damage but they are frequently preferred because the higher atomic number of germanium gives it a higher cross-section for photoelectric absorption.

The energy which is deposited in a solid state detector is used to create electrical carriers in the form of electron–hole pairs which are collected at the electrodes. The number of carriers created is usually so large that the statistical fluctuations in the number can be disregarded. But the total charge collected is very small and the detector needs to be followed by a charge amplifier with high gain and low noise; it is the noise from this amplifier which limits the resolution of the detector, typically to ~ 3 keV.

Solid state detectors are much more expensive, weight for weight, than scintillation counters. Like scintillation counters, they record only the energy deposited in the detector and if the gamma ray escapes from the detector after suffering one or more Compton scatterings the true energy of the gamma ray is not measured. Examples of the energy loss spectra obtained from a sodium iodide scintillation crystal and a germanium solid state detector which have been exposed to the two gamma ray lines from ^{60}Co, with energies of 1.17 MeV and 1.33 MeV respectively, are shown in Fig. 4.3. The greater resolution of the germanium detector gives narrower profiles to the two lines and also reveals the presence of the two Compton edges in the energy losses of the scattered photons; but the larger volume of the sodium iodide crystal, with its greater probability of total absorption, produces a spectrum in which a greater proportion of events lies in the two peaks.

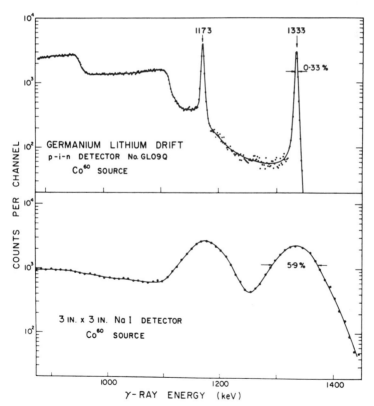

FIG. 4.3. Comparison of the energy-loss spectra recorded by a 3.5 mm deep Ge(Li) detector and a 3 in. diameter × 3 in. deep NaI scintillation crystal when exposed to 1.17 MeV and 1.33 MeV gamma rays from ^{60}Co. (From Ewan and Tavendale 1964, p. 2301.)

4.4. Passive shielding

In Chapter 3 we saw that the major problem in the design of a gamma ray telescope is the background flux of gamma rays produced by the cosmic radiation. Two techniques of shielding, commonly referred to as passive and active shielding, have been developed to cope with this problem; a passive shield consists of inert material, such as lead, which merely attenuates gamma rays from outside whilst an active shield is a second detector which not only attenuates the gamma ray flux but also vetoes any event in the central detector which is associated with a signal from the shield.

In all space experiments, whether on balloons, rockets, or satellites, the total weight of the telescope is an important factor. We shall first show that, for small detectors, the weight of the shielding decreases as the density of the shielding material increases. For simplicity we shall consider a spherical detector of radius, r, surrounded by a shield which has only a negligibly small aperture in it. To obtain the required attenuation of the background radiation in the shield the thickness of the shield must be greater than a given number of attenuation lengths. The mass attenuation coefficient for Compton scattering, which is the only interaction we need consider at these energies, is approximately the same for all materials. Let α be the necessary thickness of the shield measured in g cm^{-2} and let t be the thickness measured in cm; then

$$\alpha = \rho t,$$

where ρ is the density of the shielding material.

The outer radius, R, of the shield is given by

$$R = r + \alpha/\rho$$

and the mass, M, of the shield by

$$M = \frac{4\pi}{3} \rho (R^3 - r^3)$$

$$= \frac{4\pi}{3} (\alpha^3/\rho^2 + 3r\alpha^2/\rho + 3r^2\alpha).$$

For small detectors, $r \ll \alpha/\rho$ and $M \propto \rho^{-2}$.

The effectiveness of a passive shield is always much less than that calculated on the basis of a straightforward attenuation of atmospheric gamma rays, the difference being due to radiation created in the shield itself. In fact, since the low energy background radiation is in equilibrium with the high energy nucleonic component of the cosmic radiation, and since the high energy component is almost unattenuated by the shield, it might be expected that the low energy component would, to a first approximation, be unchanged. In a more detailed treatment of the problem two further factors need to be taken into account. First, the high energy nucleonic component actually increases in the outer layer of the shield because the production of secondary particles more than compensates for the attenuation of primary particles. Secondly, as we have seen in Chapter 3, the position of the maximum in the energy spectrum of the low energy gamma rays produced by Compton scattering depends on the atomic number of the medium in which the scattering occurs; in a medium with high atomic number the greater probability of photoelectric absorption moves the maximum to higher energies. The combination of these two effects in a shield composed of a heavy metal, such as lead, results in an attenuation of the background radiation at energies below ~ 1 MeV but an increase in the background intensity above this energy. This effect was demonstrated by an experiment (Hillier and Standing 1959) in which a small scintillation crystal surrounded by a lead shield was flown on a balloon. The lead shield contained a small aperture which was closed periodically during the flight by a lead plug (Fig. 4.4). The counting rates in the scintillation crystal were recorded with the aperture open and with the aperture closed. The difference between these two rates, which is a measure of the attenuation introduced by the lead plug, is shown in Fig. 4.5. It is clear that the lead plug decreases the background intensity at energies below ~ 0.6 MeV but increases the intensity above this energy.

A significant improvement in the shielding can be obtained by surrounding the passive shield with a second detector which will register a charged particle

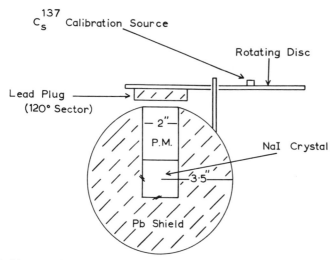

FIG. 4.4. Diagram of apparatus used to measure the effect of a passive lead shield on the background radiation at balloon altitudes. The lead plug periodically covers the aperture in the shield and the resultant changes in the counting rate of the central detector are recorded.

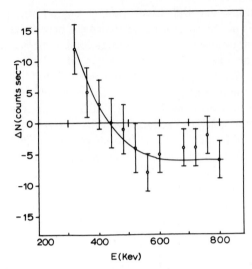

FIG. 4.5. Change in the counting rate, ΔN, observed when the lead plug was removed from the aperture of the apparatus in Fig. 4.4.

which enters the shield; the signal from this detector can then be used to veto any signal from the central detector which is associated, in time, with the passage of a charged particle into the shield. This technique eliminates the effects of background radiation created in the shield by the charged nucleonic component of the cosmic radiation. But a considerable fraction of the secondary nucleons are neutrons and these will not be registered by the outer detector. Clark, Lewin, and Smith (1968) found that the use of an anticoincidence counter of this type reduced the intensity of the background radiation at low energies by a factor of ~ 2.5.

4.5. Active shielding

The effects of background radiation produced in the shield can be greatly reduced if the whole of the shield consists of a second detector which is operated in anticoincidence with the central detector. An interaction in the shield can then be recognized if the primary particle or any of the secondary particles are charged. We have seen that the density of the shielding material should be as high as possible in order to minimize the total weight of the shield; this implies that the inorganic scintillators such as sodium iodide or caesium iodide are the most suitable materials for active shields.

The disadvantage of an active shield is the high cost, and it is often the cost and not the weight, which limits the size of the shield. Even if a very large active shield can be constructed its effectiveness will be limited because there

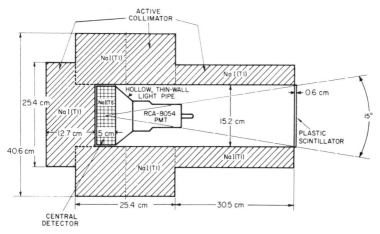

FIG. 4.6. Diagram of a low energy gamma ray telescope, built at Rice University, and consisting of a sodium iodide scintillation counter surrounded by an active shield. (From Walraven *et al.* 1975, p. 503.) © (1975) The American Astronomical Society.

are two sources of background radiation which it cannot eliminate; the first source is interactions, such as inelastic neutron scattering, where no charged particles are involved and the second is induced radioactivity where the radiation is emitted some time after the interaction occurs.

Figure 4.6 is a diagram of a scintillation detector with active shielding which was constructed at Rice University for experiments at balloon altitudes. The background counting rate in the central detector recorded during a balloon flight (Walraven, Hall, Meegan, Coleman, Shelton, and Haymes 1975) is shown in Fig. 4.7. The counting rate in a similar detector which was unshielded is also shown, this data being derived from measurements by Peterson, Schwartz, and Ling (1973). A comparison of the two rates shows that the active shielding of the Rice telescope reduced the background by a factor of ~5 at ~1 MeV, although part of this decrease is really a reduction in the efficiency of the shielded detector because gamma rays which Compton scatter in the central detector and then interact in the shield are vetoed.

4.6. Unshielded detectors

In an actively shielded detector the central detector and the shield are often constructed from similar material. We have seen that considerations of either cost or weight will limit the quantity of material which is available for a particular experiment and we shall now consider what should be the optimum division of this material between the central detector and the shield.

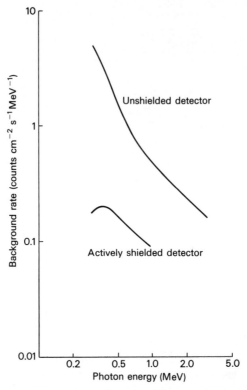

FIG. 4.7. Comparison of the background counting rates recorded at balloon altitudes by an unshielded scintillation counter and by the Rice University telescope shown in Fig. 4.6.

For simplicity we shall take the central detector to be a sphere of radius r and the shield to be a spherical shell with outer radius R. We shall assume that the aperture in the shield is sufficiently small to be ignored. In Section 3.4 we saw that the minimum flux, $s(\varepsilon)$, which a telescope can detect depends on its area A, its efficiency $\eta(\varepsilon)$, and the intensity of the background radiation $b(\varepsilon)$, according to the relationship

$$s(\varepsilon) \propto \sqrt{\{b(\varepsilon)/A\eta(\varepsilon)\}}. \tag{4.1}$$

We shall assume that $\eta(\varepsilon) \sim 1$ and that the background radiation inside the shield is predominantly gamma rays which have penetrated the shield.

We wish to examine how the value of $s(\varepsilon)$ depends on the choice of r, for a given value of R (that is, for a given total quantity of scintillator). Now

$$A = \pi r^2$$

and

$$b(\varepsilon) = b_0(\varepsilon) \exp[-\mu(\varepsilon)\{R - r\}],$$

where $b_0(\varepsilon)$ is the intensity of the background radiation outside the shield and $\mu(\varepsilon)$ is the attenuation coefficient for gamma rays in the scintillator. By substituting these expressions in eqn 4.1 and differentiating that equation with respect to ε, we find that the minimum value of $s(\varepsilon)$ occurs when

$$r = 2/\mu(\varepsilon).$$

For low energy gamma rays, $\mu(\varepsilon)$ decreases monotonically with photon energy, and there will be some energy, ε_c, at which

$$r = 2/\mu(\varepsilon_c) = R.$$

At this energy, and above, the greatest sensitivity is achieved by dispensing with the shield and using all the available scintillator to construct the largest possible detector. Clearly the geometry which has been considered in this discussion would not be used in practice, but a similar conclusion is reached if detailed calculations are made for more realistic designs.

The total mass of scintillator, M, is given by

$$M = \frac{4\pi}{3}\rho R^3$$

where ρ is the density, and in Fig. 4.8 we have plotted ε_c against M for sodium iodide and for plastic scintillator. These curves show, for example, that if the total weight of sodium iodide is limited to 10 kg then an active shield is of no advantage for measurements of photons with energies above ~ 0.9 MeV.

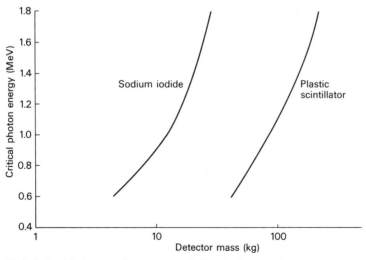

FIG. 4.8. Relationship between the mass of available scintillator and the photon energy above which active shielding is of no advantage.

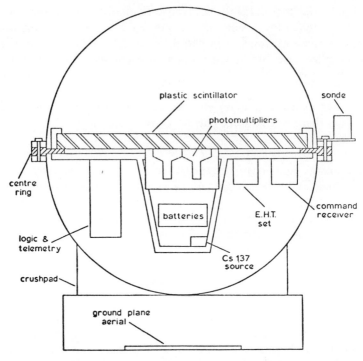

FIG. 4.9. Diagram of an unshielded scintillation counter built at University of Bristol to make measurements on the gamma ray emission from pulsars.

Figure 4.9 shows a diagram of an unshielded detector constructed at the University of Bristol to study the radiation from pulsars at photon energies of ~ 1 MeV (Sale 1970). Plastic scintillator, with a mass of 50 kg, was used in this experiment and from Fig. 4.8 we see that an unshielded detector was the appropriate design. The detector had a collecting area of 0.9 m^2, which was a factor of ~ 50 greater than that of the Rice telescope shown in Fig. 4.6, and it is clear that this more than compensates for any reduction in background intensity which could be expected from an active shield. A major disadvantage of the unshielded design is the very high counting rate recorded by the detector and the very small signal-to-background ratios expected from sources, which means that great care must be taken to avoid small systematic variations of background rate due, for example, to changes in azimuth or changes in balloon altitude.

4.7. Collimators

Since it is difficult to construct a detector for low energy gamma rays which is inherently directional in its sensitivity, measurements on localized sources require some form of collimation to distinguish between the parallel, or nearly parallel, beams of radiation from sources and the much larger diffuse flux of background radiation.

Several factors need to be considered when designing the angular response of the collimator. First, there is the need to make background measurements in a direction as near as possible to that of the source in order to minimize errors due to anisotropies in the background flux; an aperture with an opening angle of a few degrees is adequate for this purpose. Secondly, it is essential that there should be not more than one source in the aperture at any one time. This requirement becomes serious only when a telescope is capable of detecting a large number of sources in the sky. A third possible consideration is the need to make an accurate measurement of the position or the angular diameter of the source. The accuracy required in this case is usually of the order of arc minutes or even arc seconds and this demands a sophisticated collimator and guidance system for the telescope.

When an active or passive shield is used around a detector it may also be made to serve as a collimator simply by extending it in the forward direction, but if it is to provide an opening angle of a few degrees the collimator is excessively long when it is constructed in the form of a simple cylinder. The collimator is shorter, and more convenient, if it is constructed from slats as shown in Fig. 4.10(a); collimation in the perpendicular direction can be provided by a second set of slats, the two sets forming a honeycomb structure.

Let us assume that the slats are completely opaque to the radiation which is being measured. The angular response of such a collimator is triangular, the full width, θ_0, at half height being given by

$$\tan \theta_0 = y/2h,$$

where y is the distance between the slats and h is the height of the collimator.

The number of slats, n, required to cover a detector of linear dimensions, D, is given by

$$n = \frac{D}{x + y},$$

where x is the thickness of the slats. Since a fraction $x/(x + y)$ of the detector is obscured by the slats it is essential that

$$x \ll y.$$

If the slats are of length l and are constructed from material of density ρ, the total mass, M, of the collimator is given by

$$M = nxhl\rho \sim \rho Dl \tan \theta_0,$$

and we see that the total weight does not depend on the number of slats.

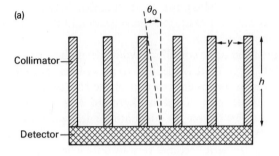

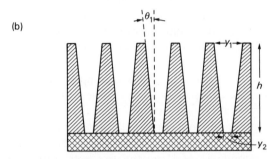

FIG. 4.10. Diagrams showing cross-sections of two forms of slat collimators: (a) slats with a simple rectangular cross-section which produce a triangular response function; (b) slats with a trapezoidal cross-section which produce a response function which is constant over an angle θ_1 about the forward direction.

The triangular response of this form of collimator is inconvenient if an accurate measurement of the intensity of radiation from a source is required because there is no direction in which the sensitivity does not vary with angle. This implies that, despite the large width of the angular response, the orientation of the collimator must be known to high accuracy. This disadvantage may be overcome by making the slats trapezoidal in cross-section as shown in Fig. 4.10(b). The angular response of this collimator is trapezoidal and the response is constant over an angle $\pm\theta_1$ about the forward direction where

$$\tan \theta_1 = (y_1 - y_2)/h.$$

The disadvantage of this design is that a larger fraction of the detector is permanently obscured by the collimator.

In some applications a more convenient form of collimator is the modulation collimator (Mertz 1966) which was first developed for measurements in X-ray astronomy. In its simplest form this consists of two grids of

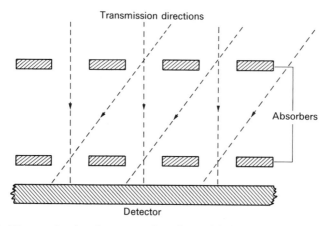

FIG. 4.11. Diagram showing the cross-section of a modulation collimator formed from two layers of parallel absorbers.

absorbers placed in front of the detector, as shown in Fig. 4.11. If we imagine the geometrical shadow of one grid on the other we see that such a system possesses several directions of maximum transmission interleaved with directions of minimum transmission. When the response pattern of the collimator is moved across the sky the signal at the detector from a localized source is modulated, the modulation being different for sources with different positions in the sky. In this way several sources can be studied at the same time; in fact an image of the whole sky can be obtained by taking the appropriate transform of the signal at the detector, the form of the transform depending on the response function of the collimator and the motion of the pattern across the sky.

A common design for this type of collimator is the rotation modulation collimator in which the two grids are rotated about an axis which is perpendicular to the planes of the grids. The signal from a source on the axis of rotation is unmodulated, but the signal from an off-axis source is modulated at a frequency which depends on the angle between the source direction and the rotation axis. This form of modulation collimator is particularly convenient for detectors in satellites because the satellites are often required to spin for stabilization purposes.

A development of the modulation collimator which has superior image-forming characteristics is the coded-mask (Dicke 1968). As developed for X-ray astronomy this consists of an opaque mask with random holes in it, behind which is a position-sensitive detector. A point source in a given direction in the sky produces a characteristic shadow on the detector and an image of the whole sky can be obtained from a convolution of the aperture function of the mask with the signal distribution at the detector. Position-sensitive detectors are less easy to construct in low energy gamma ray astronomy and an alternative approach is to use a simple detector with a second mask in front of it; an image of the sky is then derived from the signal received at the detector as the two masks are moved relative to each other.

Modulation collimators and coded-masks, with their image-forming properties, may be seen as attempts to overcome the lack of mirrors and lenses for use at gamma ray wavelengths, but they are still far from achieving the arc-second angular resolution which is obtained with only a modest optical telescope.

References

Clark, G. W., Lewin, W. H. G., and Smith, W. B. (1968). *Astrophys. J.* **151**, 21.
Dicke, R. H. (1968). *Astrophys. J.* **153**, L101.
Evans, R. D. (1975). *The atomic nucleus.* McGraw-Hill, New York.
Ewan, G. T. and Tavendale, A. J. (1964). *Can. J. Phys.* **42**, 2286.
Hillier, R. R. and Standing, K. G. (1959). University of Bristol preprint.
Mertz, L. (1966). *Transformations in optics.* John Wiley, New York.
Peterson, L. E., Jacobson, A. S., Pelling, R. M., and Schwartz, D. A. (1968). *Can. J. Phys. Suppl.* **48**, S437.
Sale, R. G. (1970). Ph.D. Thesis, University of Bristol, p. 56.
Walraven, G. D., Hall, R. D., Meegan, C. A., Coleman, P. L., Shelton, D. H., and Haymes, R. C. (1975). *Astrophys. J.* **202**, 502.

5

COMPTON TELESCOPES

5.1. Introduction

The angular distributions of the secondary photon and electron which emerge from a Compton scattering become narrower as the energy of the primary photon increases. When the energy of the primary photon is greater than ~ 3 MeV it is feasible to use the directions of motion of the secondary particles to define the direction of motion of the primary photon. A gamma ray detector which is based on this principle is known as a Compton telescope. Two forms of Compton telescope are possible, one relying on the detection of the secondary photon and the other relying on the detection of the secondary electron.

5.2. Compton scattering

Consider a Compton interaction in which a photon with energy $h\nu_0$ is scattered through an angle θ and emerges with an energy $h\nu_1$; the electron is ejected with kinetic energy E at an angle ϕ to the initial direction of motion of the photon.

From considerations of the conservation of energy and momentum it can be shown that

$$\nu_1 = \nu_0/\{1 + \alpha(1 - \cos\theta)\} \tag{5.1}$$

and

$$E = h\nu_0\alpha(1 + \cos\theta)/\{1 + \alpha(1 - \cos\theta)\}, \tag{5.2}$$

where

$$\alpha = h\nu_0/mc^2.$$

The relationship between the angles θ and ϕ is given by

$$\cot\phi = (1 + \alpha)\tan(\theta/2). \tag{5.3}$$

The differential cross-sections for Compton scattering have been derived by Klein and Nishina (1928) and only the results of the calculations will be given here. For photons scattered through an angle θ it is important to distinguish between the cross-section $d\sigma(\theta)/d\theta$, which refers to the number of photons per unit angle of scatter at an angle θ, and the cross-section $d\sigma(\theta)/d\Omega$ which refers to the number of photons per unit solid angle at an angle θ. The relationship between the two differential cross-sections is

$$\frac{\mathrm{d}\sigma(\theta)}{\mathrm{d}\theta} = \frac{\mathrm{d}\sigma(\theta)}{\mathrm{d}\Omega} 2\pi \sin\theta \, \mathrm{d}\theta. \tag{5.4}$$

The result of the Klein–Nishina calculation of the differential cross-section for the scattering of unpolarized photons is

$$\frac{\mathrm{d}\sigma(\theta)}{\mathrm{d}\Omega} = \frac{r_0^2}{2}\left(\frac{v_1}{v_0}\right)^2\left(\frac{v_0}{v_1} + \frac{v_1}{v_0} - \sin^2\theta\right), \tag{5.5}$$

where

$$r_0 = e^2/mc^2.$$

The differential cross-section $\mathrm{d}_e\sigma(\phi)/\mathrm{d}\Omega$ for the ejection of an electron at an angle ϕ may be derived from eqn 5.3 and eqn 5.5. For each photon which is scattered through an angle between θ and $\theta + \mathrm{d}\theta$ there will be an electron ejected at an angle between ϕ and $\phi + \mathrm{d}\phi$, so we may write

$$\frac{\mathrm{d}_e\sigma(\phi)}{\mathrm{d}\Omega} 2\pi \sin\phi \, \mathrm{d}\phi = \frac{\mathrm{d}\sigma(\theta)}{\mathrm{d}\Omega} 2\pi \sin\theta \, \mathrm{d}\theta.$$

Using this relationship we find

$$\frac{\mathrm{d}_e\sigma(\phi)}{\mathrm{d}\Omega} = \frac{r_0^2}{2}\left(1 - \frac{E}{hv_0}\right)^2 \left\{\left(1 + \frac{E}{hv_0}\right)^{-1} + \left(1 + \frac{E}{hv_0}\right) - \frac{4\cot^2\phi(1 + \alpha)^2}{[(1 + \alpha)^2 + \cot^2\phi]}\right\}$$

$$\times \frac{4(1 + \alpha)^2 \cot\phi \, \mathrm{cosec}^3\,\phi}{[(1 + \alpha)^2 + \cot^2\phi]^2}. \tag{5.6}$$

Figure 5.1 shows $\mathrm{d}\sigma(\theta)/\mathrm{d}\Omega$ for six different values of the primary photon energy. Figure 5.2 gives the corresponding distributions of $\mathrm{d}\sigma(\theta)/\mathrm{d}\theta$; note that in the forward direction, whereas $\mathrm{d}\sigma(\theta)/\mathrm{d}\Omega$ is a maximum, $\mathrm{d}\sigma(\theta)/\mathrm{d}\theta$ falls to zero because of the factor of $\sin\theta$ in eqn 5.4. Figure 5.3 shows the distributions of $\mathrm{d}_e\sigma(\phi)/\mathrm{d}\phi$ for the scattered electrons.

5.3. Multiple scattering of electrons

The electron which is ejected in Compton scattering is often of low energy and suffers severe multiple scattering when passing through material. This scattering is a serious problem if the direction of emission of the electron is to be used to determine the direction of motion of the primary photon.

The multiple scattering is the result of very many small deflections in the electrostatic fields of the nuclei in the material through which the electron is moving. An electron with velocity v which passes within a distance b of a nucleus with atomic number Z suffers a deflection θ which is given by (Fermi 1949)

$$\theta \sim \frac{2Ze^2}{bm}\frac{\sqrt{(1 - v^2/c^2)}}{v^2}.$$

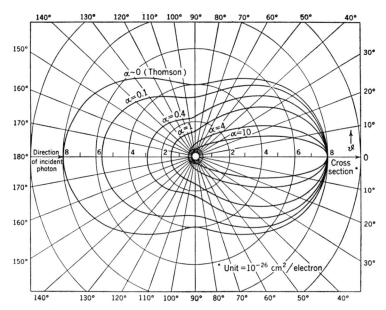

FIG. 5.1. Differential cross-sections, $d\sigma(\theta)/d\Omega$, for the production of secondary photons from Compton scattering. Curves are shown for six different values of primary photon energy. (From Davisson and Evans 1952, p. 83.)

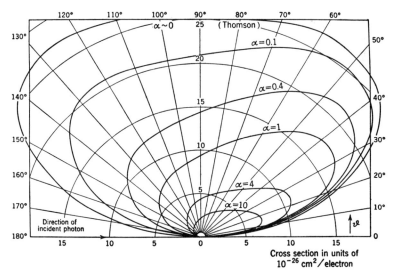

FIG. 5.2. Differential cross-sections, $d\sigma(\theta)/d\theta$, for the production of secondary photons from Compton scattering. (From Davisson and Evans 1952, p. 83.)

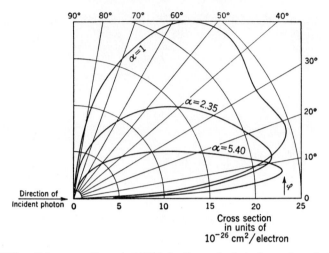

FIG. 5.3. Differential cross-sections, $d\sigma_e(\phi)/d\phi$, for the production of secondary electrons from Compton scattering. (From Davisson and Evans 1952, p. 106.)

The directions of the individual deflections are uncorrelated so, if Θ is the resultant of a large number, n, of small deflections, then

$$\Theta^2 = \sum_{i=1}^{i=n} \theta_i^2.$$

Consider an electron which passes through a thickness d of material which contains N nuclei per cm^3. The electron undergoes collisions with a range of impact parameters and the resultant deflection is given by

$$\Theta^2 = 2\pi Nd \int_{b_{\min}}^{b_{\max}} \frac{4Z^2 e^4}{bm^2} \frac{(1 - v^2/c^2)}{v^4} \, db$$

i.e. $$\Theta^2 = \frac{8\pi NdZ^2 e^4}{m^2} \frac{(1 - v^2/c^2)}{v^4} \ln(b_{\max}/b_{\min}). \qquad (5.7)$$

The minimum value of the impact parameter is set by the finite radius of the nucleus and the maximum value by the screening of the nuclear field by atomic electrons. Hence it can be shown (Heitler 1954) that

$$b_{\max}/b_{\min} \sim (183/Z^{1/3})^2.$$

For fast electrons we may put

$$\frac{mv^2}{\sqrt{(1 - v^2/c^2)}} \sim E$$

where E is the total energy of the electron. Equation 5.7 can then be rewritten as

$$\Theta^2 = \frac{16\pi N Z^2 e^4 d}{E^2} \ln\left(\frac{183}{Z^{1/3}}\right).$$

The interactions which give rise to multiple scattering are also those responsible for the production of bremsstrahlung which was discussed in Section 2.6. It is therefore possible to express the multiple scattering in terms of the radiation length, X_0, which was defined in eqn 2.43 and the result is

$$\Theta^2 = \frac{4\pi \hbar m^2 c^5}{e^2} \frac{d}{X_0} \frac{1}{E^2} \tag{5.8}$$

or

$$\Theta^2 = \frac{d}{X_0}\left(\frac{E^*}{E}\right)^2 \quad \text{where} \quad E^* \sim 21 \text{ MeV.} \tag{5.9}$$

5.4. The design of a Compton telescope using detection of the scattered photon

5.4.1. *Basic configuration*

The basic configuration of a Compton telescope which relies on the detection of scattered photons is shown in Fig. 5.4. A primary photon which suffers a

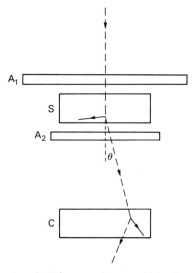

FIG. 5.4. Basic configuration of a Compton telescope which relies on the detection of scattered photons: S, scatterer; C, collector; A_1, A_2, anticoincidence detectors.

Compton scattering in the scatterer S is detected a second time in the collector C. The detectors A_1, A_2 are operated in anticoincidence with S and C and ensure that the events which are recorded are not due to penetrating charged particles.

5.4.2. *The design of the scatterer*

Consider a primary photon which is initially moving along the axis of the telescope. The probability that the photon suffers one, and only one, Compton scattering in the scatterer is

$$P_1 = \int_0^{x_S} \mu_C \exp(-\mu_C x) \exp(\mu_C x - \mu_C x_S) \, dx,$$

where μ_C is the attenuation coefficient for Compton scattering and x_S is the thickness of the scatterer. The probability P_1 is a maximum when

$$x_S = \mu_C^{-1}.$$

The telescope will accept the scattered photon only if the angle of scatter, θ, is less than θ_0, where θ_0 is the angle which the collector subtends at the scatterer. The probability that θ is less than θ_0 is

$$P_2 = \frac{\displaystyle\int_0^{\theta_0} (d\sigma/d\theta)2\pi \sin\theta \, d\theta}{\displaystyle\int_0^{\pi} (d\sigma/d\theta)2\pi \sin\theta \, d\theta},$$

where $(d\sigma/d\theta)$ is the angular distribution of the scattered photons.

Let us assume that the angular distribution of the background photons incident on the scatterer is, to a first approximation, isotropic. The angular distribution of the scattered background photons will also be isotropic, no matter what the form of the angular distribution given by eqn 5.5. The probability that a scattered background photon is accepted by the telescope is then

$$P_2' = \tfrac{1}{2}(1 - \cos\theta_0).$$

In Chapter 3 we saw that the sensitivity of a measurement on a discrete source is proportional to $N_S/\sqrt{N_B}$ where N_S is the total number of primary photons, and N_B is the total number of background photons, detected by the telescope. For a Compton telescope, $N_S/\sqrt{N_B}$ is proportional to $P_2\sqrt{(P_1/P_2')}$ and this quantity is plotted against θ_0 in Fig. 5.5 for three different values of the primary photon energy. In considering these graphs it should be noted that a single, total-absorption detector, of the type described in Chapter 4, would have $P_1 \sim P_2 \sim P_2' \sim 1$, and it would seem that Compton telescopes have no advantage over single detectors. But the assumption of an isotropic background flux is not always valid, particularly in experiments on satellites, and Compton telescopes should achieve significantly higher sensitivities. Figure 5.5 does show that Compton telescopes are not suitable for measurements at photon energies below ~1 MeV.

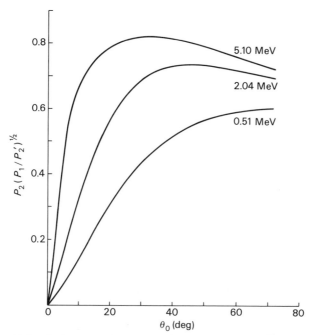

FIG. 5.5. Sensitivity of a Compton telescope, which relies on the detection of scattered photons, as a function of the acceptance angle, θ_0, of the telescope.

5.5. The design of a Compton telescope using detection of the secondary electron

5.5.1. *Basic configuration*

A design for a Compton telescope which relies on the detection of secondary electrons is shown in Fig. 5.6. There is no anticoincidence detector below the scatterer since the electron which emerges from the scatterer must be allowed to pass freely into the collector.

5.5.2. *The design of the scatterer*

As in the previous design the probability that a primary photon suffers at least one Compton scattering in the scatterer is

$$P_1 = \{1 - \exp(-\mu_C x_s)\},$$

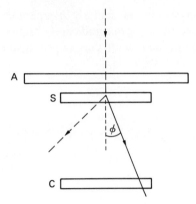

FIG. 5.6. Basic configuration of a Compton telescope which relies on the detection of secondary electrons.

where x_S is the thickness of the scatterer. A high efficiency requires a large value of x_S, but the multiple scattering of the secondary electrons also increases with x_S. In fact, the angular distribution of the electrons which emerge from a thick scatterer is determined almost entirely by multiple scattering and depends very little on the angular distribution of the electrons ejected from the Compton scatterings.

The effects of the multiple scattering of the secondary electrons can be demonstrated by a simple calculation in which it is assumed that the Compton scattering always takes place half way through the scatterer so that each secondary electron has to pass through a thickness $\frac{1}{2}x_S$ before emerging from the detector. The multiple scattering which takes place in this material leads to an angular distribution given by

$$n(\phi) = n_0 \exp(-\phi^2/4\phi_S^2),$$

where

$$\phi_S = \frac{E^*}{E} \sqrt{\left(\frac{x_S}{2X_0}\right)},$$

E^* and X_0 being defined in Section 5.3. Materials with low atomic number should be used for the scatter because they have relatively large values of the mass attenuation coefficient for Compton scattering combined with low multiple scattering of the electrons.

The probability that an electron emerges from the scatterer within the acceptance angle, θ_0, of the telescope is given by

$$P_2 = \frac{\displaystyle\int_0^{\theta_0} n(\phi)2\pi \sin\phi \, d\phi}{\displaystyle\int_0^{\pi} n(\phi)2\pi \sin\phi \, d\phi}.$$

Assuming, as before, that the angular distribution of the background photons is isotropic, the secondary electrons which leave the detector following the interaction of these photons will also be isotropic and the probability that they will be accepted by the collector is given by

$$P'_2 = \tfrac{1}{2}(1 - \cos \theta_0).$$

The number of primary photons, N_S, detected in an experiment is proportional to $(P_1 P_2)$ whilst the number of background photons detected, N_B, is proportional to $(P_1 P'_2)$. The sensitivity of the telescope depends on the ratio $N_S/\sqrt{N_B}$, or $P_2\sqrt{(P_1/P'_2)}$, and in Fig. 5.7 this quantity is plotted against θ_0, for photons with an energy of 5.1 MeV; calculations have been made for two different thicknesses of the scatterer and it is clear that, although the thickness is not critical, the acceptance angle of the telescope must be matched to the thickness chosen.

A comparison of the curves in Figs. 5.5 and 5.6 shows that the two designs of Compton telescopes can achieve comparable sensitivities. The telescope which relies on the detection of the secondary electrons has the advantage that the angular distribution of the electrons is narrower than that of the secondary photons, as can be seen from a comparison of Figs. 5.2 and 5.3, but the scatterer in this design must be thin and relatively inefficient if the secondary electron is not to suffer undue multiple scattering. This design does have the additional advantage that the collector may be of low mass because it is required to detect only charged particles and not photons.

So far we have considered the problem of optimizing the thickness of a

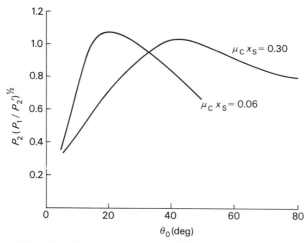

FIG. 5.7. Sensitivity of a Compton telescope, which relies on the detection of secondary electrons, plotted as a function of the acceptance angle for two different values of the scatterer thickness.

scatterer which has a given area. However, in many experiments it is the weight of the detector, and not its lateral dimensions, which is the limiting factor and in these cases the optimization procedure is different. For a scatterer of a given mass, M_S, the number of photons which undergo a Compton scattering is given by

$$n \propto \mu_C x_S A \propto \mu_C M_S \qquad (5.10)$$

and is independent of the thickness x_S.

If the thickness of the scatterer is decreased, and at the same time the area is increased to keep the mass constant, the angular resolution of the telescope can be improved and the background counting rate reduced without decreasing the number of primary photons detected. A limit to the angular resolution is set by the angular distribution of the electrons ejected from Compton scattering and there is no advantage in decreasing the thickness of the scatterer beyond the point where the multiple scattering ceases to dominate the angular distribution of the electrons leaving the scatterer.

When the thickness of the scatterer is very small the masses of the collector and the anticoincidence detector cannot be ignored. The total mass of the three detectors is given by

$$M \propto A(x_S + x_C + x_A)$$

and eqn 5.10 should be rewritten as

$$n \propto \mu_C x_S A \propto \mu_C M x_S / (x_S + x_C + x_A).$$

We see that the number of primary photons detected is independent of x_S only when

$$x_S > x_C + x_A. \qquad (5.11)$$

One technique which overcomes the limit set by eqn 5.11 is to divide the scatterer into several layers as shown in Fig. 5.8. In this design the signatures

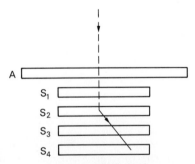

FIG. 5.8. Design for a Compton telescope in which the detectors serve as both scatterers and collectors.

which correspond to the Compton scattering of a photon are

$$(\bar{A}S_1S_2), (\bar{A}\bar{S}_1S_2S_3), (\bar{A}\bar{S}_1\bar{S}_2S_3S_4).$$

Thus each of the detectors S_2, S_3, S_4 serves as both a scatterer and a collector. It is possible to have the thickness, x_S, of each of the individual detectors sufficiently low to avoid excessive multiple scattering of the secondary electrons and yet still to comply with condition

$$x_{S_1} + x_{S_2} + x_{S_3} + x_{S_4} > x_A + x_C.$$

The principle behind this design of Compton telescope is similar to that used in the spark chamber telescopes which are discussed in the next chapter.

5.6. An example of a Compton telescope designed for use at balloon altitudes

Figure 5.9 shows the design of a Compton telescope built at the Max Planck Institut, Munich (Schonfelder, Graml, and Penningsfeld 1980). The directional sensitivity of the telescope was achieved by combining measurements of the energy and direction of motion of the scattered photon with a measurement of the energy of the secondary electron. The scatterer consisted of sixteen independent cells of liquid scintillator, each 15 cm × 15 cm × 15 cm thick; the collector consisted of thirty-two cells of sodium iodide scintillator each 15 cm × 15 cm × 7.5 cm thick. The two arrays were separated by a distance of 120 cm and each was surrounded by an anticoincidence shield. A Compton scatter was characterized by a signal from a cell in the scatterer in coincidence with one from a cell in the collector. The energy of the secondary electron was assumed to be the energy deposited in the liquid scintillator cell and the energy of the secondary photon to be the energy deposited in the sodium iodide cell.

The discussion of the basic principles of a Compton telescope in Section 5.4 assumed that only the direction of motion of the secondary photon could be measured. In that case the angular resolution of the telescope was determined entirely by the angular distribution of the secondary photons; it was possible to say only that the primary photon must have been moving in a direction within an angle, θ_S, of the direction of the secondary photon where θ_S is the width of the appropriate polar diagram in Fig. 5.1. Using the Munich telescope it was possible to place a further restriction on the direction of motion of the primary photon by deriving the angle of scatter of the photon, θ, from the two energy measurements. Rearranging eqn 5.1 we get

$$\cos \theta = 1 - \frac{mc^2}{h\nu_1} + \frac{mc^2}{E + mc^2}.$$

When θ is known the direction of motion of the primary photon is restricted to an annulus of radius θ about the direction of the scattered photon. The

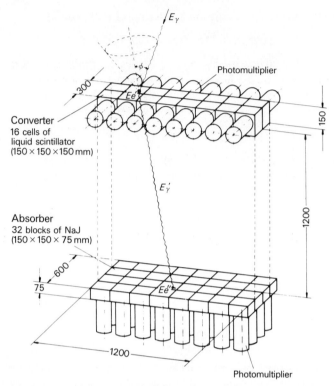

FIG. 5.9. Diagram of a Compton telescope built at the Max Planck Institut, Munich. Both the converter and the absorber were surrounded by thin anticoincidence counters of plastic scintillator which are not shown in the diagram. The dimensions are in millimetres. (From Schonfelder *et al.* 1980, p. 351.) © (1980) The American Astronomical Society.

width of the annulus depends on the precision of the energy measurements; with the Munich telescope, when a 3 MeV photon was scattered through an angle of 20° the annulus was ~1° wide.

This improvement in the angular resolution of the telescope reduced the effective background signal from atmospheric gamma rays. Further reductions in the background were achieved by accurate timing of the signals from the two arrays so that downward-moving and upward-moving photons could be distinguished and by using pulse shape measurements in the liquid scintillator to discriminate against events produced by low energy neutrons. A Monte Carlo calculation was used to derive the efficiency of the telescope and the result for 2 MeV photons is shown in Fig. 5.10.

It is interesting to compare the performance of this telescope with that of the single sodium iodide detector used by Peterson (1966). The relevant characteristics of the two instruments are given in Table 5.1. The minimum

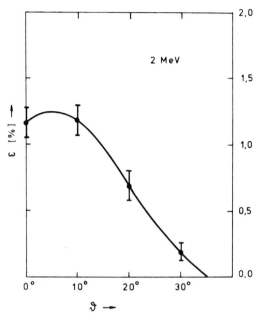

FIG. 5.10. Efficiency of the Compton telescope shown in Fig. 5.9 as function of the angle, θ, of the direction of the primary photon from the telescope axis. (From Schonfelder *et al.* 1980, p. 352.) © (1980) The American Astronomical Society.

flux which can be detected by a telescope was shown in Section 3.4 to be proportional to $\sqrt{(b/A\eta^2)}$. The data given in Table 5.1 indicates that the Compton telescope was about ten times more sensitive than the single detector; but the increase in sensitivity was due largely to the greater collecting area of the Compton telescope and if the same quantity of sodium iodide had been used in the two instruments their sensitivities would have been similar. This illustrates the very great difficulties encountered when attempting to improve the performances of low energy gamma ray telescopes.

Table 5.1. A comparison of the performance of a Compton telescope with that of a single, unshielded detector

	Compton telescope (Schonfelder et al. 1980)	Single detector (Peterson 1966)
Background counting rate per unit area in energy interval 1–10 MeV, b	2.2×10^{-4} counts cm^{-2} s^{-1}	1.1 counts cm^{-2} s^{-1}
Collecting area, A	3600 cm^2	45 cm^2
Efficiency, η	$\sim 10^{-2}$	~ 0.7

References

Davisson, C. M. and Evans, R. D. (1952). *Rev. mod. Phys.* **24**, 79.
Fermi, E. (1949). *Nuclear physics.* University of Chicago Press, Chicago, Il.
Heitler, W. (1954). *The quantum theory of radiation.* Oxford University Press, London.
Klein, O. and Nishina, Y. (1928). *Z. Phys.* **52**, 853.
Peterson, L. E. (1966). *J. geophys. Res.* **71**, 5778.
Schonfelder, V., Graml, F., and Penningsfeld, F.-P. (1980). *Astrophys. J.* **240**, 350.

6

SPARK CHAMBER TELESCOPES FOR HIGH ENERGY GAMMA RAYS

6.1. Introduction

The design of telescopes for low energy gamma rays is dictated by the characteristics of Compton scattering which is the only process through which these photons can interact with matter. The poor correlation between the direction of motion of the photon and that of the secondary electron in this interaction makes it difficult to construct a low energy gamma ray telescope with good intrinsic directional sensitivity. But, at photon energies above a few MeV the experimental situation changes because pair production, rather than Compton scattering, becomes the most probable interaction mechanism and in this process the direction of motion of the secondary electrons is much more closely aligned to that of the primary photon. Furthermore the secondary electrons, having greater kinetic energies, can penetrate considerable thicknesses of material without suffering undue multiple scattering. This means that spark chambers can be used to delineate the tracks of the secondary electrons; at lower energies the use of this form of detector is precluded because the secondary electrons are unable to penetrate the material of the spark chamber and its triggering detectors.

6.2. Pair production

In this interaction the energy of the incident photon is converted into the total energies of an electron and a positron,

$$hv \rightarrow e^+ + e^-.$$

To conserve both energy and momentum the interaction must take place in the field of a third particle which is usually that of an atomic nucleus.

The differential cross section, in cm^2 per nucleus, for the creation of an electron with kinetic energy T and a positron with kinetic energy ($hv - 2mc^2 - T$) is given by (Evans 1955)

$$\frac{d\sigma}{dT} = \frac{\sigma_0 Z^2 P}{hv - 2mc^2},$$ (6.1)

where

$$\sigma_0 = \frac{1}{137} \frac{e^4}{m^2 c^4}$$

and P is a dimensionless function of hv and T whose value varies from 0 when $hv = 2mc^2$ to ~ 20 when $hv \to \infty$. Figure 6.1 shows a plot of P against $\{T/(hv - 2mc^2)\}$ for various values of hv. The total cross-section can be derived by integrating eqn 6.1 over all values of T. This gives

$$\sigma_P = \sigma_0 Z^2 \int_0^1 P \, d(T/(hv - 2mc^2))$$

$$= \sigma_0 Z^2 \bar{P}, \tag{6.2}$$

where $\bar{P}$ is a suitably averaged value of P. For very high energy photons, and taking into account the screening of the target nuclei by atomic electrons, the value of $\bar{P}$ can be shown (Heitler 1954) to be

$$\bar{P} \sim \tfrac{28}{9} \ln(183 Z^{-1/3}) - \tfrac{2}{27}. \tag{6.3}$$

The mean distance which a photon travels before it suffers pair production is given by

$$\lambda_P = (N\sigma)^{-1},$$

where N is the number of target nuclei per unit volume. The process of pair production is related, at a fundamental level, to that of bremsstrahlung and the mean free path for pair production is related to the radiation length X_0, which was discussed in Section 2.6, by

$$\lambda_P = \tfrac{9}{7} X_0. \tag{6.4}$$

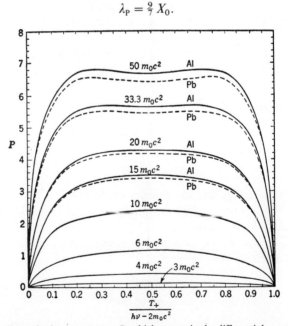

FIG. 6.1. The dimensionless parameter, P, which occurs in the differential cross-section for pair production, as a function of $(E - mc^2)/(hv - 2mc^2)$. (From Davisson and Evans 1952, p. 92.)

6.3. The basic design of a spark chamber telescope

A spark chamber consists of a number of parallel metal plates in a chamber filled with a gas, which is usually a mixture of neon and argon. Alternate plates are permanently connected to earth potential; the interleaving plates are also normally at earth potential but a large voltage pulse is applied to these plates immediately after the passage of a selected charged particle through the chamber. The charged particle leaves a trail of ion pairs in the gas and sparks occur between the plates along the path of the particle. The high voltage cannot be permanently applied to the alternate plates in the chamber because there is always a tendency for breakdown to occur, even in the absence of a trail of ion pairs in the gas, from irregularities in the surfaces of the plates. It is therefore essential to provide the chamber with triggering detectors which are designed to respond to the charged particles to be recorded; the output from these detectors is used to initiate the high voltage pulse on the plates. This need for a system of triggering detectors limits the use of spark chambers to the detection of photons with energies greater than ~ 10 MeV.

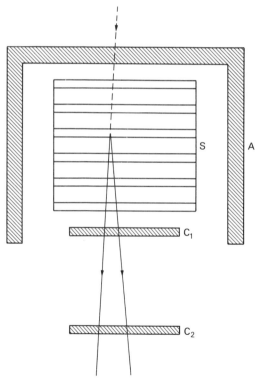

FIG. 6.2. Diagram showing the basic design of a spark chamber with plates (S), anticoincidence shield (A), and triggering detectors (C_1 and C_2).

A schematic diagram of a spark chamber telescope is shown in Fig. 6.2. A primary gamma ray, which can pass undetected through the anticoincidence counter A, may materialize into an electron pair in one of the plates of the spark chamber S. If the secondary electrons pass through the triggering detectors C_1 and C_2, a high voltage pulse is immediately applied to the plates in the chamber. The sparks which occur between the plates are recorded either by photography or by an electronic technique.

A spark chamber telescope possesses two characteristic angles, a field of view α and an angular resolution ϕ. A gamma ray can be recorded only if its initial direction of motion lies within an angle α with respect to the axis of the telescope, the value of α being determined by the geometry of the triggering detectors. Once recorded, the direction of the gamma ray can be inferred from the positions of the sparks in the chamber and the result will lie within an angle ϕ to the true direction, the value of ϕ depending on the construction of the chamber and the accuracy of the spark recording system. In general,

$$\phi \ll \alpha.$$

6.4. The angular resolution of a spark chamber telescope

In most designs of spark chamber telescopes the angular resolution is limited by the multiple scattering of the secondary electrons in the plates of the chamber. A detailed discussion of the effects of multiple scattering has been given by Pinkau (1972); the simplified treatment given here will illustrate the basic features of the problem.

Consider a photon which is initially moving along the axis of the telescope and which materializes into an electron pair midway through a plate of the chamber as shown in Fig. 6.3. The energy, ε_0, of the photon is shared between the electron and the positron according to the partition function shown in Fig. 6.1. If we assume the partition function is rectangular then the mean energy of the more energetic secondary particle is $\frac{3}{4}\varepsilon_0$ and the mean energy of the less energetic particle is $\frac{1}{4}\varepsilon_0$. We shall assume that the more energetic particle can be recognized and is used to determine the direction of motion of the photon. The electron leaves the bottom face of the plate in which it has been created and moves at an angle θ_1 to the axis of the telescope, where

$$\theta_1 \sim k\sqrt{(x/2)}.$$

The constant, k, is given by eqn 5.9. The point at which the electron leaves the plate is at a distance y_1 from the axis of the telescope, where (Fermi 1949)

$$y_1 \sim \tfrac{1}{3}k(x/2)^{3/2}.$$

The electron then traverses the gap between the plates, suffering negligible

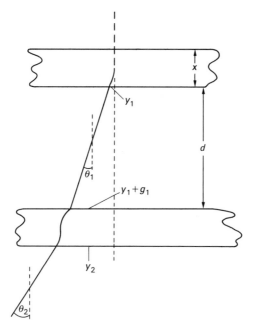

FIG. 6.3. Diagram illustrating the error due to multiple scattering in a measurement of the direction of an electron track.

multiple scattering in the gas. It strikes the second plate at a point $(y_1 + g_1)$ from the axis of the telescope where

$$g_1 = \theta_1 d.$$

The multiple scattering suffered by the electron in the second plate is uncorrelated with that in the first plate and to find the total effect we must add the two components in quadrature. The electron therefore leaves the second plate at a distance y_2 from the axis of the telescope where

$$y_2^2 = (y_1 + g_1)^2 + \tfrac{1}{3}k^2 x^3.$$

The electron is then moving at an angle θ_2 with respect to the axis where

$$\theta_2 \sim k\sqrt{(3x/2)},$$

and it will strike the third plate at a distance $(y_2 + g_2)$ from the axis where

$$g_2 = \theta_2 d.$$

We shall assume that a spark develops from the mid-point of the ion trail in each gap in the chamber. The spark in the first gap will therefore be at a distance

$$\Delta_1 = y_1 + \tfrac{1}{2}\theta_1 d$$

from the axis and the spark in the second gap will be at a distance

$$\Delta_2 = y_2 + \tfrac{1}{2}\theta_2 d$$

from the axis. From the positions of these two sparks we infer that the direction of motion of the electron, and therefore that of the primary photon, makes an angle $\Delta\theta$ to the axis of the telescope where

$$\Delta\theta = \frac{\Delta_2 - \Delta_1}{x + d}. \tag{6.5}$$

This error in the measurement of the direction of motion of the primary photon determines the angular resolution of the telescope.

6.5. Background events in a spark chamber telescope

The function of the triggering system of the telescope is to ensure that a high voltage pulse is applied to the plates as soon as possible after the materialization of a gamma ray in the chamber. The geometry of the detectors which constitute the triggering system thus defines the field of view of the telescope. The recording of the precise positions of the sparks in the chamber is one of the major technical problems in the operation of a spark chamber telescope and it is essential that the number of events to be recorded should be kept as low as possible. This may be achieved by restricting the field of view and by ensuring that particles other than gamma rays have a very low probability of triggering the telescope. There are several classes of events which, if the triggering system is not well designed, may mimic the materialization of a gamma ray in the chamber.

(a) The passage of a fast charged particle through the telescope. This type of event should be rejected by the anticoincidence detector but the flux of charged particles is at least three orders of magnitude greater than the flux of gamma rays and the efficiency of the anticoincidence detector needs to be very high to reduce this form of background to an acceptable level.

(b) A nuclear disintegration in the chamber produced by the interaction of a high energy neutron. The flux of secondary neutrons in the atmosphere is greater than the flux of gamma rays of similar energy and the neutrons, being uncharged, are not rejected by the anticoincidence counter. If a high energy neutron suffers a nuclear interaction in the chamber, one or more of the secondary charged particles emerging from the interaction may trigger the telescope. The sensitivity of the telescope to this type of event can be reduced by using Cerenkov counters rather than scintillation counters in the triggering system of the telescope; the velocity threshold of the Cerenkov detectors can be chosen so that the detectors respond to the relativisitic electrons from gamma rays but not to the slower secondary particles from nuclear interactions.

(c) The decay of a π-meson in the chamber. An upward-moving π-meson may enter the chamber without passing through the anticoincidence detector. If it comes to rest in the chamber it will then decay into a μ-meson which in turn will decay into a relativistic electron. This electron will be indistinguishable from an electron produced in the materialization of a gamma ray. We shall make a simple calculation to show that this type of event forms a significant proportion of the background triggering rate.

Deep in the earth's atmosphere the number of π-mesons which come to rest in the chamber will be approximately equal to the number created, in nuclear disintegrations, in the chamber; near the top of the atmosphere the number of stopping π-mesons will be approximately half this number because the downward-moving component will be almost absent. Consider a spark chamber with area A cm^2 and total thickness t g cm^{-2}. Let n be the flux of high energy nucleons incident on the chamber and let the interaction length of the nucleons in the chamber be λ g cm^{-2}. Then the rate of nuclear reactions in the chamber is given by

$$r_i = nAt/\lambda.$$

Powell, Fowler, and Perkins (1959) have shown that, on the average, approximately one π-meson is created in each nuclear collision of a cosmic ray nucleon. So, from the argument given above, the rate of π-mesons stopping in the chamber will be given by

$$r_\pi \sim \frac{1}{2} r_i \sim \frac{1}{2} \frac{nAt}{\lambda}.$$

To be accepted by the triggering system of the telescope the π-meson must enter the chamber through the solid angle Ω' which is not covered by the anticoincidence detector, and the secondary electron must leave through the acceptance solid angle Ω of the triggering detectors. Thus the triggering rate due to this type of event is

$$r_{t\pi} \sim nA \frac{t}{\lambda} \frac{\Omega'}{4\pi} \frac{\Omega}{4\pi}.$$

For a typical telescope, $\Omega'/4\pi \sim 0.03$, $\Omega/4\pi \sim 0.1$ and $t/\lambda \sim 0.05$ so that $r_{t\pi} \sim 2 \times 10^{-4} nA$. The number of triggers due to charged particles is $r_{tp} \sim (1 - \eta)nA$ where $(1 - \eta)$ is the inefficiency of the anticoincidence shield, so the triggers from π-mesons stopping in the chamber will be significant if the inefficiency of the anticoincidence shield is less than $\sim 10^{-4}$.

6.6. The triggering system of a spark chamber

Consider a gamma ray which is initially moving along the axis of the telescope and which materializes in the upper plate of the chamber. Let the chamber consist of N plates each with a thickness x g cm^{-2}. The mean square angle of scattering suffered by the electron before it leaves the chamber is given by

$$\langle\theta\rangle^2 \sim k^2 N x, \tag{6.6}$$

where the constant k is given by eqn 5.9. The acceptance angle, α, of the triggering counters must be large enough to accept the scattered electrons which implies that

$$\alpha^2 > k^2 N x. \tag{6.7}$$

However, the rate of background events which are recorded increases with α, and too large an acceptance angle presents difficulties with data handling.

To demonstrate the problem of data handling with a spark chamber telescope we shall make a simple calculation, assuming that all the background events are due to atmospheric gamma rays which have an intensity of b_0 photons cm^{-2} s^{-1} sr^{-1}. A telescope with a sensitive area A, an efficiency η, and an acceptance angle α will record N_B background events in a time T where

$$N_B = b_0 A \eta T \frac{(1 - \cos \alpha)}{2}$$

$$\sim b_0 A \eta T \alpha^2 / 2.$$

When the data from a spark chamber are analysed to give the intensity from a particular source, the only background events which are relevant are those within an angle ϕ with respect to the source's direction, where ϕ is the angular resolution of the telescope. In Section 6.4 we considered the effects of multiple scattering of the electrons on the angular resolution; the result given in eqn 6.5 may be approximated, when $d \gg x$, to

$$\phi \sim k\sqrt{x}. \tag{6.8}$$

In a measurement on a source the number of background events, n_B, is therefore given by

$$n_B \sim b_0 A \eta T \frac{\phi^2}{2}.$$

Since $\alpha^2 \sim N\phi^2$, we see that $N_B \sim N n_B$.

The minimum flux which can be detected by a telescope was shown in Section 3.4 to be given by

$$S \propto \sqrt{\left(\frac{b}{A T \eta^2}\right)},$$

where b is the background rate recorded by the telescope, i.e. $b = \eta b_0$. In the case of a spark chamber telescope we find that

$$S \propto \frac{\alpha^2}{\sqrt{(NN_B)}}.$$

Thus, for high sensitivity, a telescope should be capable of recording a large number of events from a chamber which has many plates and a small field of view; the product NN_B is a measure of the data-handling capacity of the telescope.

6.7. The recording of spark positions in the chamber

The angular resolution of a spark chamber telescope depends on the accuracy with which the initial directions of the secondary electrons can be measured. We considered in Section 6.4 the limitations on the angular resolution set by multiple scattering of the electrons and it is important that errors introduced in recording the spark positions should not further degrade the angular resolution. In a typical experiment the angular resolution is $\sim 3°$ which implies that, with a plate separation of 1 cm, the spark positions must be recorded with an accuracy to ~ 0.3 mm. This accuracy must be maintained over the entire volume of the chamber and the largest chambers have up to 30 plates, each with an area approaching 1 m^2 so that the number of potential spark positions may be as large as $\sim 3 \times 10^8$. A number of different techniques have been developed to meet this problem, including photography, sonic recording, and the use of wires to locate the sparks.

Photographic recording of the sparks can be used only when the telescope is recovered after the observations and so it is not suitable for experiments in satellites. A further disadvantage is the need for a subsequent scanning of the developed film, which is usually done manually, although this is partially offset by the large amount of redundant information recorded on the film which can be used to determine whether the chamber is operating satisfactorily.

Most of the spark chambers which have been built for gamma ray astronomy have used electrodes in the form of wire grids to locate the spark positions. Alternate grids are arranged in orthogonal directions so that the signals from two grids are sufficient to define a spark position. A number of techniques have been developed to locate the wires to which the discharges occur. One method is to place a small ferrite bead around each wire and to allow the current in the spark to change the state of magnetization in the bead; the states of all the beads in the array can then be read out at a convenient time after the event.

6.8. An example of a spark chamber telescope designed for a satellite experiment

The telescope built for the COS-B satellite experiment will be used to illustrate the design principles which have been considered in this chapter. The telescope was designed to detect and measure the directions of gamma rays with energies above 30 MeV and also to provide a measurement of photon energies up to ~1000 MeV.

Figure 6.4 shows the design of the telescope. The spark chamber (SC) had a sensitive area of 24 cm × 24 cm and contained 16 gaps, each of dimensions 1.5 cm. Each gap was bounded by two grids of parallel wires, the wires in the two grids being arranged orthogonally. Between each gap was a tungsten sheet 0.042 radiation lengths thick. The chamber was filled with neon at a pressure of 2 bar with 0.6% ethane added as a quenching agent.

The energy of a gamma ray was measured by the total absorption counter (E) which consisted of a caesium iodide scintillation crystal 25 cm in diameter and 4.7 radiation lengths thick. A thin plastic scintillator (D) monitored the cascades from the higher-energy gamma rays which leaked out the bottom of detector E.

The spark chamber was triggered by coincident signals from the Cerenkov counter (C), the scintillation counter (B2), and the total absorption counter

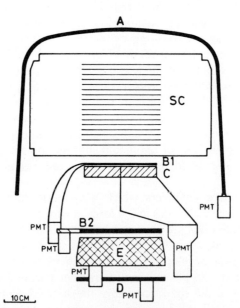

FIG. 6.4. Schematic diagram of the spark chamber telescope on the COS B experiment. (From Bennett *et al.* 1974, p. 323.)

(E). The Cerenkov counter (C) and the scintillation counter (B2) were each subdivided into four quadrants; by using the quadrants separately the field of view of the telescope could be reduced by a factor of 4. The Cerenkov counter was designed to have a low sensitivity to upward-moving particles.

The satellite was launched into a highly eccentric orbit with an apogee of $\sim 95\,000$ km. In such an orbit the satellite spends most of its time at large distances from the earth where the intensity of background radiation from the atmosphere is small, and it also passes quickly through the radiation belts where the fluxes of ionizing particles are sufficiently high to require the experiment to be switched off.

References

Bennett, K., Burger, J. J., Gorisse, M., Mayer-Hasselwander, H. A., Pfeffermann, E., Shukla, P. G., Stiglitz, R., Swanenburg, B. N., Taylor, B. G., and Wills, R. D. (1974). *Proc. 9th ESLAB Symp., ESRO SP-106*, p. 323.

Davisson, C. M. and Evans, R. D. (1952). *Rev. mod. Phys.* **24**, 79.

Evans, R. D. (1955). *The atomic nucleus*. McGraw-Hill, New York.

Fermi, E. (1949). *Nuclear Physics*. University of Chicago Press, Chicago, Il.

Heitler, W. (1954). *The quantum theory of radiation*. Oxford University Press, London.

Pinkau, K. (1972). *Nucl. Instrum. Methods* **104**, 517.

Powell, C. F., Fowler, P. H., and Perkins, D. H. (1959). *The study of elementary particles by the photographic method*, p. 423. Pergamon Press, Oxford.

7

THE DETECTION OF EXTENSIVE AIR
SHOWERS FROM VERY HIGH ENERGY
GAMMA RAYS

7.1. Introduction

The interaction of cosmic gamma rays with the earth's atmosphere, which is a great hindrance to measurements over large regions of the spectrum, can be turned to advantage at photon energies above $\sim 10^{13}$ eV. A primary photon at these very high energies may still suffer its first interaction close to the top of the atmosphere, but the absorption of the energy carried by the photon proceeds very slowly in the atmosphere and a significant fraction of the secondary radiation reaches ground level. The energy is propagated through the atmosphere in the form of an electromagnetic cascade; the multiple scattering of the particles in the cascade causes it to spread out over a large area and the cascade is often referred to as an extensive air shower. It is the lateral extent of the air shower, as well as its penetration through the atmosphere, which is of importance in gamma ray astronomy because it means that a detector at ground level which is well away from the path of the primary gamma ray may nevertheless record a signal from the secondary radiation. In other words the effective collecting area of a detector is the area of the shower front rather than the geometrical area of the detector.

7.2. The longitudinal development of an air shower

An air shower starts near the top of the atmosphere when a high energy gamma ray undergoes pair production, producing a relativistic electron and positron. In Section 6.2 we saw that the mean-free-path, λ_P, for pair production is given by

$$\lambda_P = \tfrac{9}{7} X_0 \qquad (7.1)$$

where X_0 is the radiation length for the medium. The relativistic electron and positron subsequently lose energy through the process of bremsstrahlung; in Section 2.6 it was shown that the rate of energy loss by bremsstrahlung could be written as

$$\left(\frac{dE}{dx}\right)_b = \frac{E}{X_0}. \qquad (7.2)$$

The development of an electromagnetic cascade consists of a succession of acts of pair production and bremsstrahlung. Equations (7.1) and (7.2) suggest that it is convenient to discuss the development of an air shower, not in terms of atmospheric depth z but in terms of a variable t where

$$t = z/X_0.$$

A simplified model of the longitudinal development of an electromagnetic cascade has been given by Heitler (1948). In a real cascade, fluctuations play an important role in the development but they are difficult to treat mathematically. The simplified model ignores fluctuations and assumes that each photon travels a distance $X_0 \ln 2$ and then undergoes pair production, whilst each electron travels the same distance and then radiates half its energy as a photon.

Initially the primary photon has an energy E_0. It travels a distance $X_0 \ln 2$ into the atmosphere and then materializes into an electron–positron pair, each particle having an energy $E_0/2$. After a further distance $X_0 \ln 2$ each particle radiates a photon with energy $E_0/4$ and continues with energy $E_0/4$. The two secondary photons then materialize into electron–positron pairs and in this way the numbers of photons and charged particles increase, whilst the energy of the individual particles decreases. The number of particles, N, at a depth $t \ln 2$ is given by

$$N(t \ln 2) \sim 2^t,$$

so that

$$N(t) \sim e^t,$$

and the energy of each particle is given by

$$E \sim E_0 \, e^{-t}.$$

The development of the cascade ceases when the energies of the individual electrons and positrons are approximately equal to the critical energy, E_c, because ionization rather than bremsstrahlung then becomes the dominant form of energy loss. The maximum number of particles is reached at a depth t_{max} where

$$E_c \sim E_0 \exp(-t_{max})$$

or

$$t_{max} \sim \ln\left(\frac{E_0}{E_c}\right).$$

In this model the number of photons is approximately equal to one-third of the total number of particles. At the maximum development the total number of particles is given by

$$N(t_{max}) \sim \frac{2}{3}\left(\frac{E_0}{E_c}\right).$$

An interesting result which can be derived from the model is that the total track length of the electrons and positrons is equal to E_0/E_c radiation lengths.

The results of more detailed calculations (Snyder 1949) on the development of electromagnetic cascades are shown in Fig. 7.1. The variation in the number of particles, N, against the distance, t, from the origin is shown for seven cascades with energies from 10^6 MeV to 10^{12} MeV. The development of a cascade is characterized by a parameter s, the number of particles being a maximum when $s = 1$. The atmosphere is more than 25 cascade units thick so cascades are fully developed before they reach ground level.

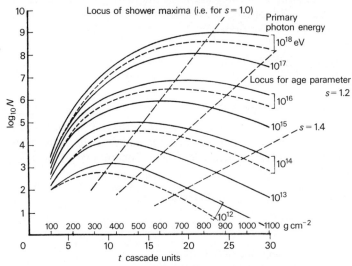

FIG. 7.1. Number of particles, N, in an electromagnetic cascade as a function of depth, t, measured in cascade units. Curves are shown for six different energies of the primary photon. (From Cranshaw 1963, p. 66.)

7.3. The lateral development of an air shower

The multiple Coulomb scattering of the charged particles in the atmosphere causes a shower to spread out as it develops. The mean square angle of scattering suffered by an electron with energy, E, as it traverses a distance, d, in a medium was derived in Section 5.3 and was shown to be

$$\Theta^2 = \frac{4\pi\hbar m^2 c^5}{e^2} \frac{d}{X_0} \frac{1}{E}.$$

The lowest-energy electrons in a cascade have an energy E_c and it is these which suffer the greatest multiple scattering. At sea level an electron with energy E_c is scattered through a distance of ~ 80 m whilst passing through one radiation length. This distance, which we shall denote by R_1, is a useful unit with which to measure the lateral extent of an air shower.

The lateral distribution of the particles in a shower was derived by Moliere (1946). This calculation was based on two diffusion equations, one for the photons and the other for the electrons and positrons, each containing terms representing scattering and the creation and loss of particles. No general analytic solution of the equations is possible but Nishimura and Kamata (1952) derived an approximate solution for showers near their maximum development. The lateral distribution of the shower particles can be represented by a density distribution, $\Delta(r')$, which expresses the number of shower particles per unit area at a distance r' from the shower axis where

$$r' = r/R_1.$$

Nishimura and Kamata showed that the density distribution is given by

$$\Delta(r') = \frac{Nf(r')}{R_1{}^2},$$

where N is the total number of particles in the shower and $f(r')$ is a function common to all showers and is shown in Fig. 7.2. From the form of $f(r')$ we see that the majority of the shower particles are contained within a distance of $0.1R_1$ from the shower axis, but that some particles occur at ten times that distance.

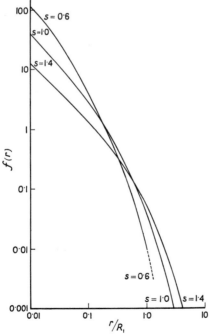

FIG. 7.2. The function, $f(r)$, which describes the lateral distribution of electrons about the shower axis. (From Cranshaw 1963, p. 68.)

7.4. The detection of air showers using arrays of charged particle detectors at ground level

A shower with sufficient energy to penetrate to ground level can be detected with an array of charged particle detectors, usually geiger counters or scintillation counters. A simple calculation will show that the detection efficiency of an array may be high even though the fraction of the ground covered by detectors may be very small. Consider an array of r detectors each with a sensitive area A. Let the mean density of shower particles, averaged over the shower front, be Δ particles per unit area. Then the mean number of shower particles recorded by each detector will be $A\Delta$. Poissonian fluctuations will cause the actual number of particles recorded to vary from detector to detector. The probability that a detector records no particle is $\exp(-A\Delta)$. The probability that all r detectors record at least one particle is $\{1 - \exp(-A\Delta)\}^r$. Thus it is possible to record showers with reasonable efficiency provided that the detectors have sensitive areas of the order of $1/\Delta$. If we put $A = 1$ m^2 then we can record showers with $\Delta \sim 1$ particle m^{-2}; this density occurs at the maximum development of showers with total energies greater than $\sim 10^7$ MeV. The separation of the detectors should be somewhat less than the diameter of the shower front which is of the order of 25 m, so we see that the detectors need cover less than 1 % of the area of the array. The direction of the shower axis can be measured by accurate timing of the arrival of the shower front at different detectors.

7.5. The detection of the Cerenkov light from air showers

An alternative method of recording an extensive air shower is to detect with photomultipliers the Cerenkov light which the shower particles emit as they travel through the atmosphere. Since the Cerenkov light in a low refractive index medium such as air is radiated only in the forward direction, a shower produces an almost parallel beam of light and the direction of the shower can be determined by mounting the photomultipliers at the focus of a parabolic mirror.

The sensitivity of this technique is limited by the intensity of the background light of the night sky. On moonless nights, at sites well away from man-made lights, the intensity of light from the night sky ϕ_b is $\sim 6 \times 10^7$ photons cm^{-2} s^{-1} sr^{-1} in the wavelength interval 4300–5500 Å. Consider a photomultiplier with a photocathode efficiency, η, and a resolving time, τ, mounted at the focus of a mirror with a collecting area, A. The Coulomb scattering of the charged particles in the shower causes the Cerenkov light to be emitted over a finite solid angle, Ω, about the shower axis. The mean number of photoelectrons emitted from the photocathode in the resolving time, τ, due to background light is given by

$$N_0 \sim \phi_b A \eta \tau \Omega.$$

In a typical experiment $A \sim 10^4$ cm^2, $\eta \sim 0.2$, $\tau \sim 20$ ns and $\Omega \sim 0.2$ sr, giving

$$N_b \sim 500 \text{ photoelectrons}$$

Jelley (1967) gives the number of Cerenkov photons emitted by the shower as $\sim 30\varepsilon$ where ε is the energy of the shower in MeV. At ground level the photons are spread over an area of $\sim 10^8$ cm^2. In the example we have chosen the number of photoelectrons emitted from the photocathode, produced by Cerenkov photons from the shower, will be given by

$$N_s \sim 3 \times 10^{-7} \varepsilon \eta A$$

$$\sim 6 \times 10^{-4} \varepsilon.$$

The signal from the shower will be detectable above the fluctuations in the background if

$$N_s > 4N_b,$$

that is

$$\varepsilon > 2 \times 10^5 \text{ MeV}.$$

We see that the Cerenkov technique typically has a lower energy threshold and it also has two other advantages. The Cerenkov light is spread over a greater area than the charged particles, which gives the detector a greater effective collecting area, and the optical system of the Cerenkov collecting system allows the direction of the shower axis to be determined without the need for accurate timing of the shower front.

7.6. An example of a ground-based system for detecting atmospheric Cerenkov light from an extensive air shower

A system for detecting extensive air showers was built around two large optical reflectors at Narrabri Observatory, New South Wales. The reflectors, which had originally been constructed for use as an intensity interferometer to measure stellar diameters, were steerable and were mounted on carriages which could be moved around a circular track with a radius of 94 m (Fig. 7.3). Each reflector was 7 m in diameter and carried a photomultiplier at its focus. The two detectors were operated in coincidence and the reflectors were orientated so that they collected light from a section of an air shower axis which was at an atmospheric depth of ~ 300 g cm^{-2} which corresponded to the maximum development of a shower with an energy of $\sim 10^5$ MeV. The system had an acceptance angle for air showers of $\sim 5 \times 10^{-4}$ sr and an effective collecting area of $\sim 1.6 \times 10^8$ cm^2. The threshold energy for the detection of gamma rays was $\sim 2 \times 10^5$ MeV.

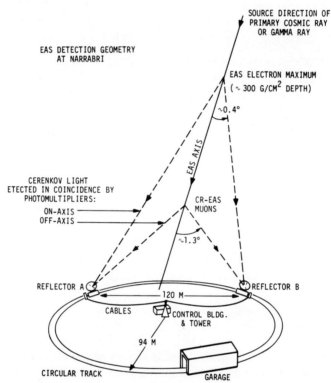

FIG. 7.3. Diagram of a detector system for air showers at Narrabri Observatory, New South Wales. (From Grindlay, Helmken, Hambury-Brown, Davis, and Allen 1975, p. 83.) © (1975) The American Astronomical Society.

References

Cranshaw, T. E. (1963). Cosmic rays. Oxford University Press, London.

Grindlay, J. E., Helmken, H. F., Hanbury Brown, R., Davis, J., and Allen, L. R. (1975). *Astrophys. J.* **201**, 82.

Heitler, W. (1948). The quantum theory of radiation. Oxford University Press, London.

Jelley, J. V. (1967). *Prog. elem. Part. cosmic ray Phys.* **9**, 41.

Moliere, G. (1946). *Cosmic radiation* (ed. W. Heisenberg). Dover, New York.

Nishimura, J. and Kamata, K. (1952). *Prog. Theor. Phys.* **7**, 185.

Snyder, H. S. (1949). *Phys. Rev.* **76**, 1563.

8

SOLAR GAMMA RAYS

8.1. The quiet sun

The bulk of the radiation emitted by the quiet sun originates in the photosphere, a surface layer only 300 km deep. The spectrum of this radiation approximates to that of a black body with a temperature of 5750 K, and consequently contributes a negligible flux to either the X-ray or the gamma ray region of the spectrum.

Above the photosphere lies the corona, a region of very low density which extends far into the solar system. The corona has a temperature of $\sim 2 \times 10^6$ K and is a copious source of soft X-rays with a spectrum which extends up to a few keV but then falls rapidly at higher energies. Gamma ray astronomy has therefore played no role in the study of the quiet sun.

8.2. Solar activity

The sun is the only star which is close enough for us to study events on its surface in detail, and a rich variety of transient phenomena, known collectively as solar activity, has been observed. Our understanding of the mechanisms of many of these phenomena is very incomplete, but the basic cause of solar activity seems to be the interaction between the solar magnetic field and the motions due to differential rotation and convection in the outer layers of the sun.

The most readily observed aspect of solar activity is the formation of sunspots. These appear as dark spots on the solar disc which last for periods ranging from days to weeks. A sunspot is a region in the photosphere where the temperature is ~ 1100 K below that of the surrounding area and the thermal emission is therefore ~ 60 per cent lower. Magnetic fields of the order of ~ 1000 G are associated with sunspots and it is the pressure exerted by the magnetic field which prevents the sunspot from collapsing under the higher thermal gas pressure of the surrounding area. The occurrence of sunspots is unpredictable but the average rates show a periodic variation with a period of 11.2 years (Fig. 8.1).

Prominences are another aspect of solar activity, which often occur in association with sunspots. A prominence is a loop of gas which extends from the photosphere far out into the corona. The gas in the loop is cooler than

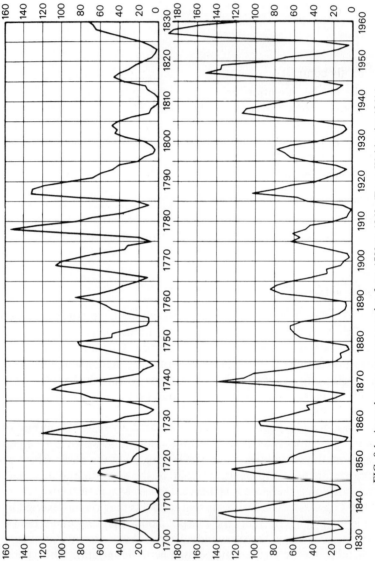

FIG. 8.1. Annual mean sunspot number from 1700 to 1960. (From Waldmeier 1961.)

that in the photosphere and the loop shows up as a dark filament when seen against the disc of the sun, but the thermal emission from the loop is greater than that from the much more tenuous corona and the prominence appears bright when seen against the corona at the limb of the sun. Prominences often occur in regions between sunspots of opposite magnetic polarity which suggests that the gas in the prominence is supported by the horizontal magnetic field in the region.

A third, and perhaps the most dramatic, aspect of solar activity is the occurrence of solar flares. Since these are frequently sources of hard X-rays, and sometimes gamma rays, they will be considered in detail in the following sections.

8.3. General characteristics of solar flares

Historically, flares were first noticed as regions near sunspots where the optical brightness, either in white light or in H_α radiation, increased by a large factor (frontispiece). Flares usually occur in association with large groups of sunspots and, like prominences, are found between sunspots of opposite magnetic polarity which suggests that the magnetic field is the source of energy for the flare. A typical example of the variation of the optical emission from a flare is shown in Fig. 8.2. The initial brightening, referred to as the explosive phase, lasts for only a few minutes and during this phase the area of the flare region increases rapidly. The explosive phase is followed by a decay which is much slower, and some optical emission is often still detectable several hours after the onset.

Solar flares have been classified into groups according to their magnitude or, to use the commonly accepted term, their importance. Table 8.1 shows how the importance is defined in terms of the area of the flare. It is now recognized that the optical emission from flares, although readily observed and classified, is only a secondary effect, the fundamental process being the acceleration of particles to high energies by sudden changes in the magnetic

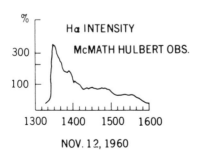

FIG. 8.2. Variation with time of the intensity of H_α radiation from a solar flare. (From Sakurai 1974, p. 106.)

Table 8.1. Relationship between the area of a flare and its importance

Importance	Area of flare (millionths of hemisphere)
1 −	< 100
1	100–250
2	250–600
3	600–1200
3 +	> 1200

field in the flare region. Several other secondary effects can be detected. For example, flares of importance 3 or higher are usually accompanied by bursts of radio emission. The radio bursts have been classified into five groups—type II, type III, type IV, type V, and microwave bursts—according to their spectra and polarization. Bursts of type II and type III are due to plasma oscillations, probably induced by the passage of energetic particles through the flare region. The characteristics of the other types of bursts indicate that they are synchrotron radiation from fast electrons; the microwave bursts are of particular interest because they occur during the explosive phase of the flare when particle acceleration is thought to be taking place.

8.4. Evidence for particle acceleration in solar flares

8.4.1. *Cosmic ray particles reaching the earth*

The first evidence that cosmic ray particles were accelerated in solar flares came from observations of variations in the intensity of secondary cosmic ray particles at the earth's surface which followed large flares on the sun. To produce a detectable signal at ground level a primary cosmic ray proton at the top of the atmosphere must have an energy greater than ~ 5 GeV. Such particles are produced only by flares of importance 3 and above, and these occur only infrequently, for example, 19 cases were observed between 1942 and 1968.

Since 1958 solar cosmic ray particles of much lower energy have been detected with instruments on balloons and satellites. These experiments detect the primary cosmic ray particles and they have shown that the chemical composition of the particles is similar to that of the photosphere and the corona. Particles with energies of a few MeV frequently accompany flares of importance 2 and above.

The frequency of flares which produce detectable fluxes of cosmic ray particles varies, as expected, through the 11-year cycle of solar activity.

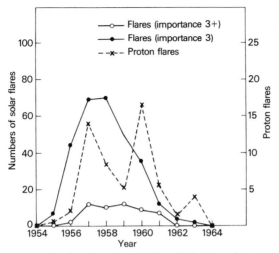

FIG. 8.3. Frequency of major flares from 1954 to 1964. (From Sakurai 1974, p. 125.)

Figure 8.3 compares the frequency of flares producing detectable cosmic ray particles with the frequency of major flares during the years from 1955 to 1963; there is some indication that the two distributions are not identical.

The flux of particles with energies above ~ 10 MeV detected at approximately 1 AU from the sun following a flare ranges from $\sim 10^{-3}$ cm^{-2} s^{-1} to 10^4 cm^{-2} s^{-1}. The energy spectrum of the particles varies from event to event. It can be expressed most conveniently in terms of the magnetic rigidity, R, where

$$R = \frac{pc}{ze},$$

p being the momentum. The rigidity spectrum generally has the form

$$\frac{\mathrm{d}J}{\mathrm{d}R} = \left(\frac{\mathrm{d}J}{\mathrm{d}R}\right)_0 \exp\left(\frac{R}{R_0}\right),$$

where $(\mathrm{d}J/\mathrm{d}R)_0$ and R_0 are functions of time and also vary from event to event. Some typical rigidity spectra are shown in Fig. 8.4.

8.4.2. Hard X-ray and microwave emission from solar flares

The X-rays which are detected from solar flares can be separated into a thermal component, which has a soft spectrum and a slow decay time, and a non-thermal component, which has a power law spectrum and which occurs only during the explosive phase of the flare. The characteristics of the soft component indicate that it is emitted by plasma in the flare region which has undergone sudden heating, probably caused by the passage of energetic

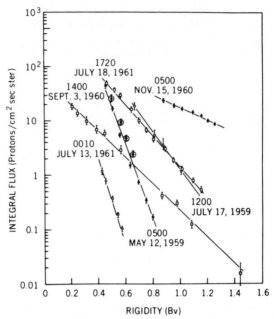

FIG. 8.4. Rigidity spectra of particles observed from six flares. (From Freier and Webber 1963, p. 1607.)

particles through the region. The non-thermal component has been inter-preted as bremsstrahlung from fast electrons and, like the microwave emission, it is evidence that particle acceleration takes place during the explosive phase of the flare.

Holt and Ramaty (1969), using data recorded during the flare of 7th July 1966, have considered the relationship between the different forms of evidence for particle acceleration in solar flares. Hard X-rays were observed with energies up to 500 keV during this flare. Figure 8.5 shows that the temporal variations of the X-ray flux correlated strongly with those of the microwave emission and this supports the argument that both forms of radiation were produced from a flux of energetic electrons.

The slope of the X-ray spectrum did not change during the decay phase of the flare which is characteristic of the radiation from a flux of relativistic electrons which are losing energy mainly through bremsstrahlung (Ginzburg and Syrovatskii 1964). Holt and Ramaty deduced that the ambient density of the medium was 10^9–10^{10} cm^{-3} and that the strength of the magnetic field in the region must have been less than 400 G for the energy loss in the form of synchrotron radiation to be negligible. The bremsstrahlung came principally from electrons with energies below 100 keV, the total number of electrons being 10^{35}–10^{37}. The particles observed directly in interplanetary space

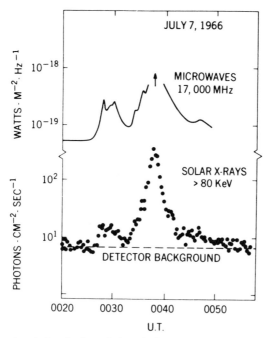

FIG. 8.5. Temporal variations in the emission of microwaves and X-rays during the solar flare
of 7th July 1966. (From Cline, Holt, and Hones 1968.)

during this event had energies above 3 MeV (Cline and McDonald 1968). If
the energy spectrum of the electrons derived from the X-ray observations is
extrapolated to higher energies the number of electrons predicted above
3 MeV is one to two orders of magnitude higher than those actually
observed. It is therefore necessary to assume either that the energy spectrum
steepened at higher energies or that the majority of the electrons did not
escape into interplanetary space.

8.5. Observations of gamma rays from solar flares

The first indication that low energy gamma rays are produced in solar flares
came on 20th March 1959 when Peterson and Winckler detected photons
with energies up to ~ 500 keV during a flare of importance 2 (Peterson and
Winckler 1959). The measurements were made with geiger counters and an
ionization chamber on a high altitude balloon; the energy resolution of the
detectors was too low to differentiate between spectral lines and a continuous
spectrum. The gamma ray burst lasted approximately 1 min at the start of the
flare and coincided with an intense microwave burst. The authors interpreted

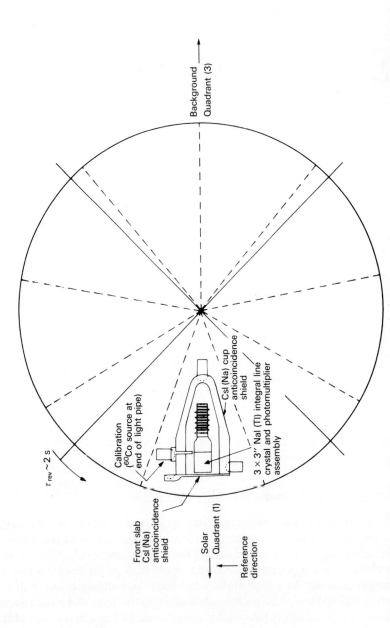

FIG. 8.6. Schematic drawing of the gamma ray spectrometer on the OSO-7 satellite. The spectrometer occupied one quadrant of the wheel of the satellite which rotated at a rate of ~0·5 s⁻¹. (From Chupp et al. 1973, p. 333.)

the gamma rays as bremsstrahlung from fast electrons and the total radiated power of $\sim 10^{24}$ ergs s^{-1} required $\sim 10^{34}$ electrons with energies up to ~ 1 MeV.

Gamma ray spectral lines were detected from solar flares in August 1972 by Chupp *et al.* (1973) using instruments on the OSO-7 satellite. The detector in this experiment was a 3 in. diameter × 3 in. thick sodium iodide crystal surrounded by an active shield of caesium iodide (Fig. 8.6). The detector was situated in the wheel of the satellite and the rotation of the satellite caused the aperture of the telescope to sweep past the sun once every ~ 2 s. The circle swept out by the aperture was divided into four quadrants, one quadrant being centred on the sun and the other three considered as background quadrants. Data were stored from the four quadrants for 90 wheel rotations before being transmitted to the ground. The satellite had a period of

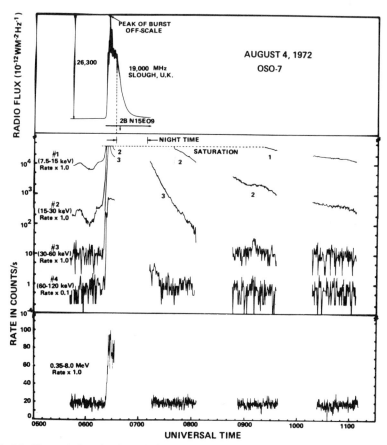

FIG. 8.7. The variations in the emission of gamma rays, X-rays, and microwave radiation during the solar flare of 4th August 1972. (From Forrest, Chupp, Suri, and Reppin 1973, p. 167.)

~93 min and for approximately half the orbit the sun was eclipsed by the earth. The data were stored in a 377-channel analyser covering the energy interval 0.3 MeV to 9.0 MeV. The instrument was calibrated twice on each orbit using gamma ray lines at 1.17 MeV and 1.33 MeV from ^{60}Co. The energy resolution was ~8 per cent at 660 keV and varied inversely as the square root of the photon energy.

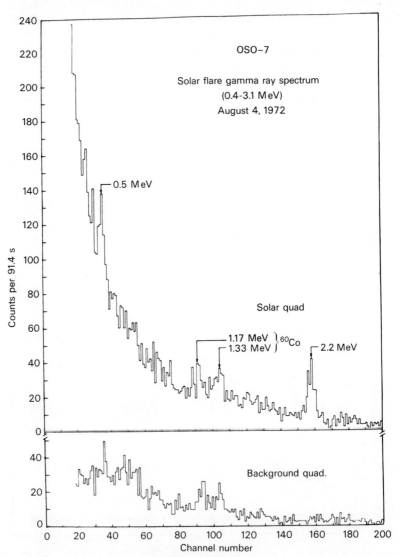

FIG. 8.8 Pulse height spectra recorded in the time interval 06.23–06.32 UT on 4th August 1972. (From Forrest *et al.* 1973, p. 169.)

Gamma rays were observed from a flare on 4th August 1972 which was classified as being of importance 3. The temporal behaviour of the counting rate in the gamma ray detector is shown in Fig. 8.7, together with that of the X-ray flux recorded in an independent detector on OSO-7 and also the variations in the microwave emission reported by Lincoln and Leighton (1972). The satellite was eclipsed by the earth at 06.32 UT; when it re-emerged from the eclipse at 07.15 UT the counting rate from the gamma ray detector had returned to the background level.

The time-averaged spectrum recorded between 06.23 UT and 06.32 UT is shown in Fig. 8.8. The continuum extended to energies above ∼ 3 MeV and there were pronounced photopeaks at 0.51 MeV and 2.2 MeV; the peaks at 1.17 MeV and 1.33 MeV were due to the calibration source. The history of the counting rates in the 0.51 MeV and 2.23 MeV photopeaks is shown in Fig. 8.9. The statistics were poor but the intensity of the gamma ray lines appeared to follow that of the continuum radiation. There were also suggestions of peaks at 4.4 MeV and 6.1 MeV; although these features were of only low statistical significance they do correspond to two spectral lines which are expected from flare regions.

Gamma rays were detected from another flare three days later on 7th

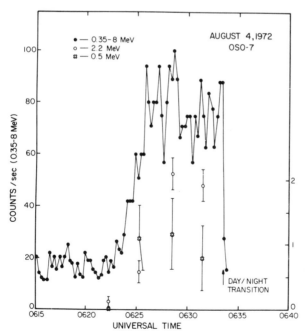

FIG. 8.9. Variation with time of the gamma ray emission during the flare on 4th August 1972. (From Chupp, Forrest, and Suri 1975, p. 341.)

August 1972. This flare was also of importance 3, but the spacecraft was behind the earth when the flare began at 14.55 UT. Approximately 40 min later the spacecraft emerged from the earth's shadow and recorded spectra showing spectral lines at 0.51 MeV and 2.2 MeV. A summary of the line intensities recorded from the two flares is given in Table 8.2.

Table 8.2. Summary of gamma ray lines observed from solar flares on 4th August 1972 and 7th August 1972

Associated flare and the time of observations	Designations and solar flux at 1 AU(photons cm^{-2} s^{-1})			
	0.5 MeV	*2.2 MeV*	*4.4 MeV*	*6.1 MeV*
3B (Hα) 4th Aug. 1972 0624–0633 UT (Before Hα Max)	$(7 \pm 1.5) \times 10^{-2}$	$(2.2 \pm 0.2) \times 10^{-1}$	$(3 \pm 1) \times 10^{-2}$	$(3 \pm 1) \times 10^{-2}$
3B (Hα) 7th Aug. 1972 1538–1547 UT (After Hα Max)	$(3.7 \pm 0.9) \times 10^{-2}$	$(4.8 \pm 1) \times 10^{-2}$	$<2 \times 10^{-2}$	$<2 \times 10^{-2}$

Gamma ray lines were also observed from a flare on 11th July 1978 (Hudson *et al.* 1980) using detectors on the HEAO-1 satellite, from a flare on 9th November 1979 (Prince *et al.* 1982) using detectors on the HEAO-3 satellite and from a flare on 7th June 1980 (Chupp *et al.* 1981) using detectors on the SMM satellite.

8.6. Interpretation of the gamma ray observations from solar flares

Ramaty, Kozlovsky, and Lingenfelter (1975) have shown how the gamma ray observations can be used to determine the physical conditions in the flare region. They assumed that the 0.51 MeV line comes from positron annihilation, that the 2.23 MeV line is produced in the capture of slow neutrons by hydrogen, and that the lines at 4.43 MeV and 6.13 MeV come from the de-excitation of ^{12}C and ^{16}O respectively.

The most copious sources of positrons are the decay of π-mesons and the decay of proton-rich radioactive isotopes; the characteristics of the most important isotopes, and the reactions in which they are produced, are shown in Table 8.3. The flux of annihilation radiation produced depends on the depth of the source region. In one model of a flare region, referred to as the thick-source model, the accelerated particles are assumed to move down-wards into the sun where they eventually lose all their energy. In the alternative thin-source model the particles escape from the sun without losing a significant fraction of their energy. It is necessary to assume some form of

Table 8.3. Positron-emitting isotopes expected to be produced in solar flares

β^+ emitter and decay mode	Maximum positron energy (MeV)	Half-life (min)	Production mode	Threshold energy (MeV)
$C^{11} \to B^{11} + \beta^+ + \nu$	0.97	20.5	C^{12} (p, pn) C^{11}	20.2
			N^{14} (p, 2p2n) C^{11}	13.1
			N^{14} (p, α) C^{11}	2.9
			O^{16} (p, 3p3n) C^{11}	28.6
$N^{13} \to C^{13} + \beta^+ + \nu$	1.19	9.96	N^{14} (p, pn) N^{13}	11.3
			O^{16} (p, 2p2n) N^{13}	5.54
$O^{14} \to N^{14} + \beta^+ + \nu$	1.86	1.18	N^{14} (p, n) O^{14}	6.4
$O^{15} \to N^{15} + \beta^+ + \nu$	1.73	2.07	O^{16} (p, pn) O^{15}	16.54

momentum spectrum for the accelerated particles and Ramaty et al. have made calculations for spectra, $n(p)$, which have the form of either an exponential,

$$n(p) = n_0 \exp\left(\frac{-p}{p_0}\right)$$

or a power law,

$$n(p) = n_0 \left(\frac{p}{p_0}\right)^{-s}$$

where p_0 and s are constants.

In the thick-source model the yield of radioactive nuclei in the source region is given by

$$q_s = n_T \int_0^\infty n(p)\, dp \int_0^x \sigma(p) \exp\left\{\frac{-(x - x')}{L}\right\} dx',$$

where n_T is the number of target nuclei per unit volume, σ is the cross-section for the production of the radioactive isotope, L is the nuclear interaction mean free path for the accelerated particles and x is the distance the particles traverse before being brought to rest by ionization losses. The corresponding expression for the thin source model is

$$\frac{dq_s}{dt} = n_T \int_0^\infty n(p)\sigma(p)v\, dp,$$

where v is the velocity of the accelerated particles. Similar equations can be written down for the production of π-mesons.

The calculated yields of positron emitters in the two models, with exponential momentum spectra, are shown in Figs. 8.10 and 8.11. We have seen in Chapter 2 that the annihilation of a free positron produces two photons, each with an energy of 0.51 MeV, whereas the annihilation of a positron after the formation of positronium has a greater probability of producing three photons with a continuous energy spectrum. In a plasma with high density and temperature, collisions quickly destroy any positronium which is formed and virtually all annihilations produce 0.51 MeV gamma rays. Ramaty and Lingenfelter (1973) have shown that positronium formation is not significant at temperatures above 10^6 K and may therefore be neglected in solar flares. The yield of 0.51 MeV gamma rays for the thick-source model is shown in Fig. 8.12 and for the thin-source model in Fig. 8.13.

The principal neutron-producing reactions in solar flare regions are given in Table 8.4 and the cross-sections for the most important of these are shown in Fig. 8.14. The cross-sections can be inserted into equations similar to those used in the calculation of the positron sources to predict the production of neutrons in a flare. As an example, Fig. 8.15 shows the neutron production predicted for an exponential momentum spectrum in the thin-source model.

The capture cross-section, σ_c, for neutrons in hydrogen is given by

$$\sigma_c \sim 2.2 \times 10^{-6} \frac{v}{c} \text{ b}$$

where v is the velocity of the neutron. This cross-section becomes significant

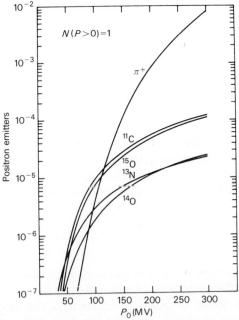

FIG. 8.10. Yield of positron emitters in the thick-source model of a solar flare with an exponential momentum spectrum of accelerated particles. (From Ramaty 1973, p. 299.)

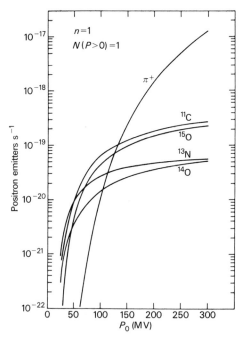

FIG. 8.11. Production rate of positron emitters in the thin-source model of a solar flare with an exponential momentum spectrum of accelerated particles. (From Ramaty 1973, p. 301.)

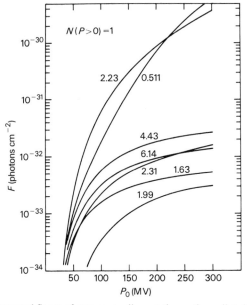

FIG. 8.12. Time-integrated fluxes of gamma ray lines at the earth predicted in the thick-source model of a solar flare with an exponential spectrum of accelerated particles. (From Ramaty 1973, p. 303.)

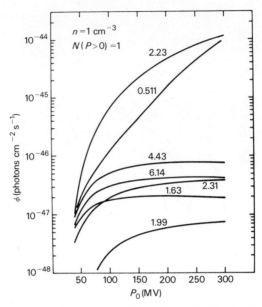

FIG. 8.13. Intensities of gamma ray lines at the earth predicted in the thin-source model of a solar flare with an exponential spectrum of accelerated particles. (From Ramaty 1973, p. 304.)

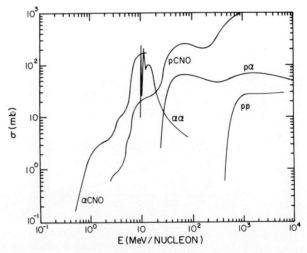

FIG. 8.14. Cross-sections for the production of neutrons. (From Ramaty *et al.* 1975, p. 347.)

Table 8.4. Principal neutron-producing reactions in solar flares

Reaction	Threshold (MeV $nucleon^{-1}$)
1. $p + {}^1\mathrm{H} \rightarrow n + p + \pi^+$	292.3
2. $p + {}^4\mathrm{He} \rightarrow {}^3\mathrm{He} + p + n + (\pi)$	25.7
$\rightarrow {}^2\mathrm{H} + 2p + n + (\pi)$	32.6
$\rightarrow 3p + 2n + (\pi)$	35.4
3. $p + {}^{12}\mathrm{C} \rightarrow n + \cdots$	19.6
$p + {}^{13}\mathrm{C} \rightarrow n + \cdots$	3.2
4. $p + {}^{14}\mathrm{N} \rightarrow n + \cdots$	6.3
5. $p + {}^{16}\mathrm{O} \rightarrow n + \cdots$	16.6
$p + {}^{18}\mathrm{O} \rightarrow n + \cdots$	2.5
6. $p + {}^{20}\mathrm{Ne} \rightarrow n + \cdots$	15.9
7. $p + {}^{56}\mathrm{Fe} \rightarrow n + \cdots$	5.5
8. $\alpha + {}^4\mathrm{He} \rightarrow {}^7\mathrm{Be} + n$	9.5
9. $\alpha + {}^{12}\mathrm{C} \rightarrow n + \cdots$	2.8
$\alpha + {}^{13}\mathrm{C} \rightarrow n + \cdots$	*
10. $\alpha + {}^{14}\mathrm{N} \rightarrow n + \cdots$	1.5
11. $\alpha + {}^{16}\mathrm{O} \rightarrow n + \cdots$	3.8
$\alpha + {}^{18}\mathrm{O} \rightarrow n + \cdots$	0.21
12. $\alpha + {}^{20}\mathrm{Ne} \rightarrow n + \cdots$	2.16
$\alpha + {}^{22}\mathrm{Ne} \rightarrow n + \cdots$	0.15
13. $\alpha + {}^{56}\mathrm{Fe} \rightarrow n + \cdots$	1.37
14. $\alpha + {}^{25}\mathrm{Mg} \rightarrow n + \cdots$	*
$\alpha + {}^{26}\mathrm{Mg} \rightarrow n + \cdots$	*
15. $\alpha + {}^{29}\mathrm{Si} \rightarrow n + \cdots$	0.43

* exoergic.

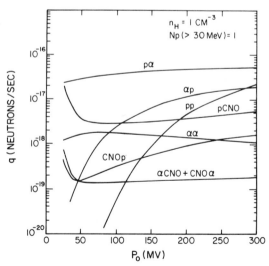

FIG. 8.15. Production rate of neutrons in a thin-source model of a solar flare with an exponential spectrum of accelerated particles. (From Ramaty *et al.* 1975, p. 349.)

only at very low velocities and the neutrons must be slowed down in collisions before they can be captured. The total neutron–proton cross-section is approximately constant at ~ 20 b for neutron energies from 1 eV to 10^5 eV and, using the data given by Allen (1973) on the densities in the chromosphere, we find that an upward-moving neutron from a flare region usually escapes from the sun without suffering a single collision. Consequently it is only the downward-moving neutrons which are moderated and then captured by hydrogen nuclei. Even these neutrons must lose their energy quickly or they will decay before they are captured. The capture time, t_c, is given by

$$t_c = (v n \sigma_c)^{-1} \sim 1.5 \times 10^{19} \, n^{-1} \text{ s},$$

and is independent of energy. The mean lifetime of a neutron before decay is 720 s so the neutron will undergo capture if

$$t_c \ll 720 \text{ s}$$

i.e. $n \gg 10^{17} \text{ cm}^{-3}.$

Densities of this magnitude occur in the lower chromosphere and a large fraction of the downward-moving neutrons can be expected to be captured by hydrogen nuclei before they decay. At these densities in the chromosphere the burden of overlying material amounts to ~ 2 g cm^{-3} which is an order of magnitude less than the attenuation length of 2.23 MeV gamma rays so most of the upward-moving photons produced in the neutron capture reactions will escape from the sun.

The emission of the lines at 0.51 MeV and 2.23 MeV should be delayed with respect to the acceleration process because the isotopes producing the positrons have lifetimes ranging from seconds to minutes and because the neutrons take periods of minutes to be thermalized and captured. By contrast the emission of the lines at 4.43 MeV and 6.13 MeV from excited nuclei should occur promptly since the lifetimes of the excited nuclei are very short. For the flare of 11th July 1978, assuming that the microwave emission coincided with the acceleration phase, the peak in the emission of the 2.23 MeV line was delayed by ~ 90 s (Hudson et al. 1980).

The intensities of the lines at 4.4 MeV and 6.1 MeV, relative to that at 2.23 MeV, depend on the momentum spectrum of the accelerated particles, since it is only the more energetic particles which can produce the neutrons required for the 2.23 MeV line. If the momentum spectrum has an exponential form the relative intensities of the lines depend only on the characteristic momentum p_0; Fig. 8.16 shows the relative intensities as a function of p_0 for the thick-source model and Fig. 8.17 shows the relative intnsities for the thin-source model. The ratios observed during the flare of 4th August 1972 are indicated as shaded regions and it appears that a value of

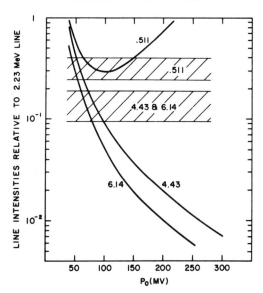

FIG. 8.16. Relative intensities of the gamma ray lines in a thick-source model of a solar flare with an exponential spectrum of accelerated particles. The relative intensities observed during the flare on 4th August 1972 are shown as shaded regions. (From Ramaty 1973, p. 308.)

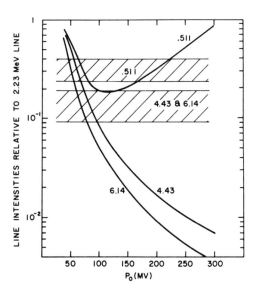

FIG. 8.17. Relative intensities of the gamma ray lines in a thin-source model of a solar flare with an exponential spectrum of accelerated particles. The relative intensities observed during the flare on 4th August 1972 are shown as shaded regions. (From Ramaty, 1973, p. 309.)

p_0 lying between 60 MV and 80 MV is required to explain the observations, irrespective of the model. This value of p_0 is consistent with the momentum spectrum of the accelerated protons as measured in the vicinity of the earth by Bostrom, Kohl, and McEntire (1972).

The number of accelerated particles detected directly at the earth, or by spacecraft in orbit around the sun, seems to be considerably smaller than the number required to produce the observed flux of gamma rays from a flare. This was particularly true for the flare on 7th June 1980, which occurred at a site on the sun such that the particles leaving the sun should have been guided to the earth by the interplanetary magnetic field. Ramaty, Lingenfelter, and Kozlovsky (1982) have shown that the number of protons with energies greater than ~ 30 MeV which escaped from the sun during this event was $\sim 10^{31}$ whereas the number required to produce the observed flux of

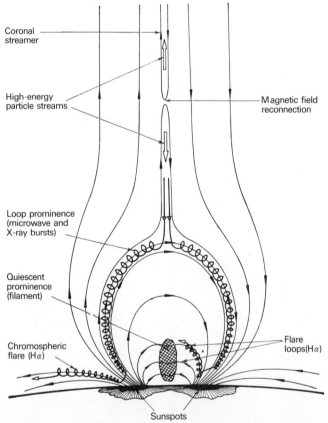

FIG. 8.18. Schematic diagram of an active solar region. An instability develops in the magnetic field above a bipolar region on the surface and reconnection of the magnetic field leads to the acceleration of beams of high energy particles. (From Strauss and Papagiannis 1971, p. 370.) © (1971) The American Astronomical Society.

2.23 MeV gamma rays was $\sim 10^{33}$. This suggests that most of the accelerated particles remained trapped in the sun and that the thick-source model of the flare was the appropriate one.

8.7. The mechanism responsible for particle acceleration in solar flares

It has long been recognized that the energy released in solar flares is probably derived from the enhanced magnetic fields which are observed in active regions on the sun. A typical magnetic field of ~ 1000 G extending over a flare region of $\sim 10^{28}$ cm^3 contains $\sim 10^{33}$ ergs, which is more than sufficient to account for the energies of all the components observed in a flare. Moreover, under certain conditions, the magnetic energy can be transformed with high efficiency into the kinetic energy of streams of fast particles which seem to be the underlying cause of the phenomena seen in flares. Several mechanisms which have been proposed would take place in the configuration of magnetic field, shown in Fig. 8.18, which lies above a bipolar region on the surface of the sun. Near the surface the magnetic field forms closed loops but at greater distances it is drawn out into approximately radial directions by the highly conducting plasma of the solar wind. On either side of the plane which bisects the bipolar region the magnetic field is parallel, but of opposite sense. Sweet (1958) argued that a plasma instability could occur in this region as a result of which the magnetic field undergoes reconnection as shown in Fig. 8.18. When this happens streams of high energy particles will be accelerated, moving both inward and outward in radial directions.

References

Allen, C. W. (1973). *Astrophysical quantities*. Athlone Press, London.

Bostrom, C. O., Kohl, J. W., and McEntire, R. W. (1972). *The solar proton flux—August 2–12, 1972*. The Johns Hopkins University Applied Physics Laboratory, Baltimore, MD.

Chupp, E. L., Forrest, D. J., Higbie, P. R., Suri, A. N., Tsai, C., and Dunphy, P. P. (1973). *Nature (London)* **241**, 333.

——, ——, Ryan, J. M., Cherry, M. L., Reppin, C., Kanbach, G., Rieger, E., Pinkau, K., Share, G. H., Kinzer, R. L., Strickman, M. S., Johnson, W. N., and Kurfess, J. D. (1981). *Astrophys. J.* **244**, L171.

——, ——, and Suri, A. N. (1975). In *Solar gamma, X-, and EUV radiation, IAU Symp. No. 68* (ed. S. R. Kane). Reidel, Dordrecht.

Cline, T. L., Holt, S. S., and Hones, E. W. (1968). *J. Geophys. Res.* **73**, 434.

—— and McDonald, F. B. (1968). *Solar Phys.* **5**, 507.

Forrest, D. J., Chupp, E. L., Suri, A. N., and Reppin, C. (1973). In *Gamma ray astrophysics*, (eds. F. W. Stecker and J. I. Trombka). NASA, Washington, DC.

Freier, P. S. and Webber, W. R. (1963). *Jr. Geophys. Res.* **68**, 1605.

Ginzburg, V. L. and Syrovatskii, S. I. (1964). *The origin of cosmic rays*. Pergamon, Oxford.

Holt, S. S. and Ramaty, R. (1969). *Solar Phys.* **8**, 119.

Hudson, H. S., Bai, T., Gruber, D. E., Matteson, J. L., Nolan, P. L., and Peterson, L. E. (1980). *Astrophys. J.* **236**, L91.
Lincoln, J. V. and Leighton, H. I. (1972). *World Data Centre A, Rep. UAG-21.* U.S. Department of Commerce, NOAA, NC.
Peterson, L. E. and Winckler, J. R. (1959). *J. Geophys. Res.* **64**, 697.
Prince, T. A., Ling, T. C., Mahoney, W. A., Riegler, G. R., and Jacobson, A. S. (1982). *Astrophys. J.* **244**, L171.
Ramaty, R. (1973). In *Gamma ray astrophysics* (ed. F. W. Stecker and J. I. Trombka). NASA, Washington, DC.
——, Kozlovsky, B., and Lingenfelter, R. E. (1975). *Space Sci. Rev.* **18**, 341.
—— and Lingenfelter, R. E. (1973). *Proc. 13th Int. Cosmic Ray Conf.*, p. 1590. University of Denver Press, Denver, CO.
——, ——, and Kozlovsky, B. (1982). *Gamma ray transients and related astrophysical phenomena.* A.I.P. Conf. Proc. (USA) no. 77, pp. 211–29 (1981). (Workshop on gamma ray transients and related astrophysical phenomena, La Jolla, CA, USA, 5–8 Aug. 1981.)
Sakurai, K. (1974). Physics of solar cosmic rays. University of Tokyo Press, Tokyo.
Strauss, F. M. and Papagiannis, M. D. (1971). *Astrophys. J.* **164**, 369.
Sweet, P. A. (1958). *Proc. I.A.U., Symp. no. 6*, p. 123. D. Reidel Publishing Co., Dordrecht.
Waldmeier, M. (1961). *The sunspot-activity in the years 1610–1960.* Schulthess, Zürich.

9

GAMMA RAYS FROM THE GALACTIC DISC

9.1. Introduction

One of the arguments which Morrison put forward in his original paper (Morrison 1958), urging the importance of gamma ray astronomy, was the information which gamma ray measurements would give on the fluxes and interactions of cosmic ray particles in remote regions of the Galaxy. In this respect, gamma ray measurements can be likened to radio measurements of synchrotron radiation which tell us about the interaction of cosmic ray electrons with the magnetic fields in the Galaxy. Before we consider the gamma ray measurements we shall therefore look at our present knowledge of the cosmic ray particles and of the distribution of gas in the Galaxy.

9.2. Cosmic ray particles in the Galaxy

9.2.1. *Cosmic ray electrons*

The energy spectrum of the cosmic ray electrons in the vicinity of the earth is shown in Fig. 9.1. At energies below ~ 1 GeV the flux of particles at the earth differs from that in interstellar space because of the modulating action of magnetic fields in the solar system, the strength of this modulation varying with the 11-year cycle of solar activity. Above ~ 5 GeV the spectrum can be expressed as

$$n(E) \sim 130 E^{-2.6} \text{ electrons m}^{-2} \text{ s}^{-1} \text{ sr}^{-1} \text{ GeV}^{-1}.$$

The interpretation of the radio emission from the Milky Way as synchrotron radiation from cosmic ray electrons shows that these particles are also present throughout the Galaxy. A representative map of the radio emission, at a frequency of 150 MHz, is shown in Fig. 9.2. There is a strong band of emission along the Galactic plane and a further enhancement towards the centre of the Galaxy. There is also some indication of weak emission from a large approximately spherical region, or halo, centred on the Galaxy. Assuming a typical interstellar magnetic field of a few microgauss the radio measurements, which range from 8 MHz to 3000 MHz, correspond to electron energies from ~ 0.5 GeV to ~ 10 GeV. The energy spectrum of the electrons can be derived from the radio measurements; the result is shown in Fig. 9.1 and at high energies where modulation in the solar system is insignificant it agrees well with the direct measurements made at the earth.

9.2.2. *Cosmic ray nuclei*

The relative abundances of the nuclei in the cosmic rays (see Table 9.1) resembles, in general, the natural abundances of elements in the universe, but the few differences are significant. For example, the much larger abundances of the light elements lithium, beryllium and boron in the cosmic rays are assumed to be the result of spallation in collisions of heavier nuclei in the interstellar gas; the abundances of these elements have been used to demonstrate that the cosmic rays are stored in the Galaxy for a period of $\sim 3 \times 10^6$ years (Shapiro and Silberberg 1970).

When calculating the production of neutral pions in collisions of cosmic ray nuclei it is usually sufficient to consider only the hydrogen and helium nuclei because of the much smaller abundances of the heavier elements. The

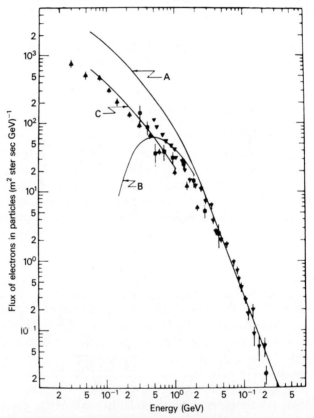

FIG. 9.1. The energy spectrum of cosmic ray electrons. The data points are direct measurements made at the earth. Curve A is the spectrum derived from measurements of the Galactic radio emission; curves B and C are calculations of the spectra expected near the earth when the Galactic spectrum is modulated by mechanisms which are rigidity dependent or velocity dependent respectively. (From Anand, Daniel and Stephens 1968, p. 26.)

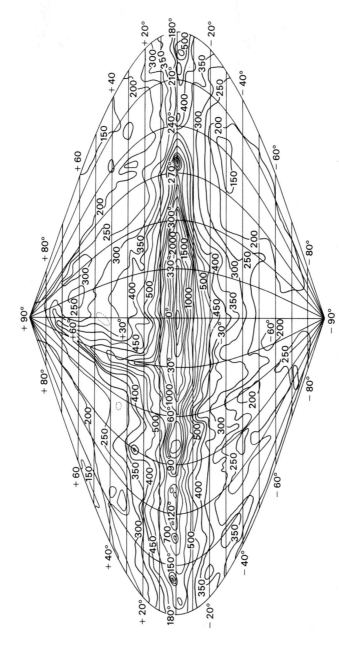

FIG. 9.2. Radio map of the sky, in Galactic coordinates, at a frequency of 150 MHz. (From Landecker and Wielebinski 1970.)

energy spectrum of the hydrogen and helium nuclei at the earth is shown in
Fig. 9.3. As with the cosmic ray electrons, the shape of the spectrum at low
energies is modified by magnetic fields in the solar system, and in Fig. 9.3
spectra are shown for three different levels of modulation. Above ~ 10 GeV
the spectrum of the hydrogen nuclei can be represented by

$$n(E) \sim 7 \times 10^4 E^{-2.6} \text{ particles m}^{-2}\text{ s}^{-1}\text{ sr}^{-1}\text{ GeV}^{-1}$$

where the energy, E, is measured in GeV. The energy spectrum of the heavier
nuclei, when expressed in terms of GeV per nucleon, has a similar shape but
with an intensity reduced by the appropriate factor from Table 9.1.

Table 9.1. Comparison of the relative abundances of the
chemical elements in the cosmic rays and in the solar
system (the two sets of abundances have been normalized
to a value of 100 at carbon)

Element		z	Cosmic rays	Solar system
H		1	26 000	270 000
He		2	3600	18 728
Li		3	18 ± 2	4.2×10^{-4}
Be	L	4	10.5 ± 1	6.9×10^{-6}
B		5	28 ± 1	3.0×10^{-3}
C		6	100	100
N	M	7	25 ± 2	31.7
O		8	91 ± 4	182
F		9	1.7 ± 0.4	2.1×10^{-2}
Ne		10	16 ± 2	29.2
Na		11	2.7 ± 0.4	0.51
Mg		12	19 ± 1	8.99
Al		13	2.8 ± 1	0.72
Si		14	14 ± 2	8.47
P	H	15	0.6 ± 0.2	8.1×10^{-2}
S		16	3 ± 0.4	4.24
Cl		17	0.5 ± 0.2	4.83×10^{-2}
A		18	1.5 ± 0.3	0.99
K		19	0.8 ± 0.2	3.6×10^{-2}
Ca		20	2.2 ± 0.5	0.611
Sc		21	0.4 ± 0.2	3.0×10^{-4}
Ti		22	1.7 ± 0.3	2.35×10^{-2}
V		23	0.7 ± 0.3	2.22×10^{-3}
Cr		24	1.5 ± 0.4	0.108
Mn	VH	25	0.9 ± 0.2	7.88×10^{-2}
Fe		26	10.8 ± 1.4	7.03
Co		27	<0.2	1.87×10^{-2}
Ni		28	0.4 ± 0.1	0.407
Cu		29		4.58×10^{-3}
Zn		30		1.05×10^{-2}
		31–35	$c. 5 \times 10^{-3}$	2.1×10^{-3}
		36–40	$c. 5 \times 10^{-4}$	9.5×10^{-4}
	VVH	41–60	$c. 5 \times 10^{-4}$	3.0×10^{-4}
		61–80	$c. 2 \times 10^{-4}$	4.7×10^{-5}
		>80	$c. 10^{-4}$	4.0×10^{-5}

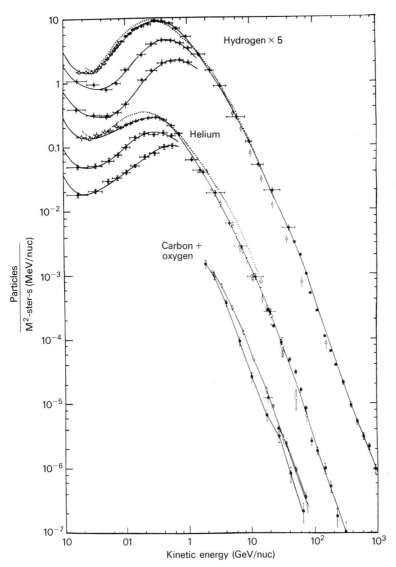

FIG. 9.3. Energy spectra of cosmic ray nuclei. The energy scale is in units of GeV per nucleon.
(From Webber and Lezniak 1974, p. 363.)

9.3. The interstellar gas

The interstellar gas, which is predominantly hydrogen, exists in the Galaxy as molecules, as neutral atoms, and as ions, and different techniques are required to detect the gas in each of these physical states.

9.3.1. *Neutral hydrogen atoms*

The density of neutral hydrogen gas in so-called HI regions can be mapped from the emission at 1420 MHz due to transitions between the hyperfine splitting of the ground state of the hydrogen atom. The neutral hydrogen is confined to a layer in the Galactic plane with a thickness of ~ 200 pc. The narrow width of the spectral line allows the radial velocity of the gas to be accurately determined and this, together with a dynamical model of the rotation of the Galaxy, can be used to draw a map of the neutral hydrogen density in the plane of the Galaxy. This map (Fig. 9.4) shows that the gas lies in a series of nearly concentric rings which resemble the spiral structure apparent in other galaxies. The average gas density within a spiral arm is ~ 1 atom cm^{-3}.

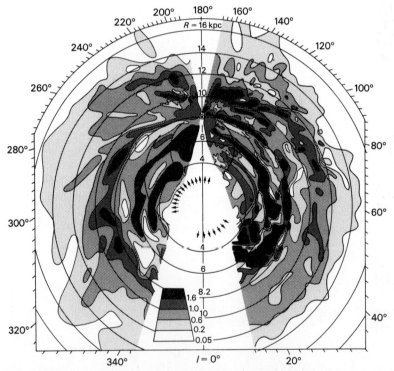

FIG. 9.4. Distribution of neutral hydrogen in the Galactic plane determined from the radio line emission at 1420 MHz. (From Oort, Kerr, and Westerhout 1958, p. 382.)

9.3.2. *Ionized hydrogen*

The detailed structure of the interstellar gas seems to be that of dense clouds lying in a more rarefied medium. When the density is sufficiently high the clouds become the sites for star formation and the ultraviolet emission from the hot young stars ionizes the hydrogen and creates an HII region. The continuum radio emission from the HII regions indicate that their densities range from 10 to 10^4 atoms cm^{-3}.

9.3.3. *Molecular hydrogen*

Cool molecular hydrogen produces no convenient spectral lines at optical or radio wavelengths which can be studied from ground-based observatories. There have been some ultraviolet measurements from spacecraft (Savage, Bohlin, Drake, and Budich 1977), but most of the evidence for interstellar molecular hydrogen is indirect and comes from studies of other interstellar molecules. For example, the dominant mechanism for the excitation of CO molecules is considered to be collisions with H_2 molecules. These theoretical arguments suggest that there may be as many as 4000 H_2 clouds in the Galaxy with densities as high as 10^4–10^5 atoms cm^{-3}. The clouds are ~ 100 pc in diameter and a large number are concentrated in a ring which extends from 4 to 8 kpc from the Galactic centre (Scoville and Solomon 1975). There is also evidence for a large cloud at the Galactic centre with a mass of $\sim 5 \times 10^7$ M. (Stark and Blitz 1978).

9.4. Measurements of gamma rays from the Galactic disc

9.4.1. *Introduction*

The first attempts to detect gamma rays with energies ~ 100 MeV were made with detectors carried by high altitude balloons. The early experiments served merely to underline the experimental problems represented by the low primary fluxes and the high intensity of locally produced background. A counter telescope on the OSO-III satellite (Kraushaar *et al.* 1972) produced evidence of gamma ray emission from the Galactic disc but there was still uncertainty about the contamination of the results with background radiation. It was widely recognized that the spark chamber was the detector most likely to succeed in these circumstances and several spark chamber telescopes were flown on balloons in the 1960s. Although some of these were large and complex (Browning, Ramsden, and Wright 1971; Frye *et al.* 1971) they produced no consistent evidence for gamma ray fluxes. It was not until the launching of two satellites, SAS-2 in 1972 and COS-B in 1975, which were devoted to gamma ray astronomy and contained spark chamber telescopes, that the first convincing measurements were made.

9.4.2. *The SAS-2 satellite experiment*

A diagram of the gamma ray telescope is shown in Fig. 9.5. The spark chamber had a sensitive area of $\sim 600 \text{ cm}^2$ and consisted of 32 modules interleaved with tungsten plates, each 0.03 radiation lengths thick. A module contained two parallel planes of wires, with a spacing of 1.27 mm, to determine the position of the sparks, alternate modules having wires in orthogonal directions. Magnetic cores threaded onto the wires were used to record the spark position. The trigger counters were a plastic scintillation counter below the sixteenth module and a Cerenkov counter below the bottom module. An anticoincidence counter completely covered the top of the assembly.

The threshold energy for the telescope was $\sim 30 \text{ MeV}$. Energy measurements could be made up to $\sim 200 \text{ MeV}$; above this energy only a lower limit to the energy could be set and therefore only integral fluxes could be measured. The angular resolution of the telescope was predicted to be $\sim 5°$ at 30 MeV, increasing to $\sim 2°$ at 1000 MeV, but in practice it seemed to be somewhat poorer.

The satellite was launched on 15th November 1972 into a low equatorial orbit with an apogee of 610 km and a perigee of 44 km. The collection of data ended on 8th June 1973 when a low voltage power supply failed. At that time only ~ 55 per cent of the sky had been examined, as is shown in Fig. 9.6. During the measurements the satellite spun at a rate of $\sim 1 \text{ rev min}^{-1}$. The spin axis drifted at a rate of a few degrees per day and the orientation of the telescope was determined to $\sim 0.3°$ by magnetometers and, independently, to $\sim 0.2°$ by star sensors.

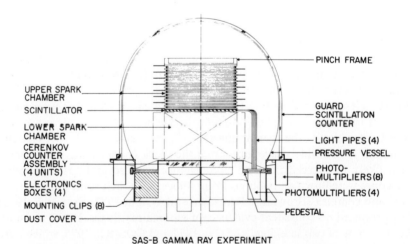

SAS-B GAMMA RAY EXPERIMENT

FIG. 9.5. Diagram of the spark chamber telescope in the SAS-2 satellite. (From Fichtel *et al.* 1975, p. 165.) © (1975) The American Astronomical Society.

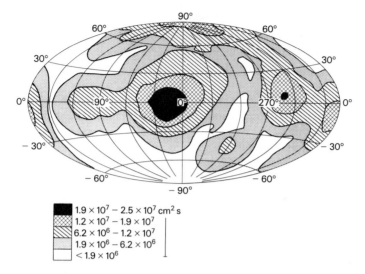

FIG. 9.6. Regions of sky observed by the SAS-2 telescope, shown in Galactic co-ordinates. The exposures are given in units of (effective area × time) for photons with an energy of 100 MeV. (From Hartman *et al.* 1979, p. 598.) © (1979) The American Astronomical Society.

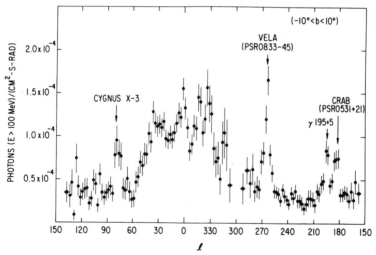

FIG. 9.7. SAS-2 measurements of the distribution of gamma ray emission ($\varepsilon > 100$ MeV) along the Galactic plane. (From Hartman *et al.* 1979, p. 599.) © (1979) The American Astronomical Society.

9.4.3. *The COS-B satellite experiment*

The COS-B gamma ray telescope has been described in detail in Section 6.7. The spark chamber had a sensitive area of 576 cm^2 and the spark positions were determined by magnetic read-out from 16 planes of wires, the wire spacing being 1.25 mm. Interleaved between the planes of wires were tungsten sheets, each 0.042 radiation lengths thick. The triggering system consisted of a scintillation counter and a directional Cerenkov counter. Below the spark chamber was an energy calorimeter consisting of a caesium iodide scintillation counter with a depth of 4.7 radiation lengths. An anticoincidence counter covered the spark chamber and the upper triggering counter.

The threshold energy for the telescope was ~ 50 MeV and the calorimeter provided crude energy measurements up to ~ 2000 MeV. The angular resolution varied from $\sim 10°$ at 50 MeV to $\sim 2°$ at 2000 MeV.

The satellite was launched on 9th August 1975 into an eccentric polar orbit with an apogee of 99, 102 km and a perigee of 346 km. The highly eccentric orbit ensured that the satellite spent the major part of its life at large distances from the earth where the flux of albedo gamma rays from the earth was small. The satellite spun at a rate of ~ 10 rev min^{-1} and gas jets were used to orientate the spin axis. The attitude of the telescope was measured to better than 0.5° using sun and earth sensors.

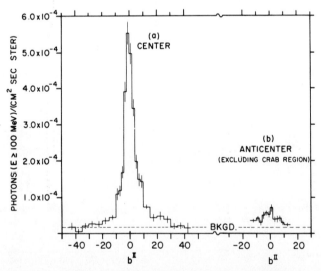

FIG. 9.8. SAS-2 measurements of the distribution of gamma ray emission ($\varepsilon > 100$ MeV) in Galactic latitude: (a) measurements in the direction of the Galactic centre; (b) measurements in the direction of the Galactic anticentre, but excluding emission from the Crab nebula. (From Fichtel *et al.* 1975, p. 170.) © (1975) The American Astronomical Society.

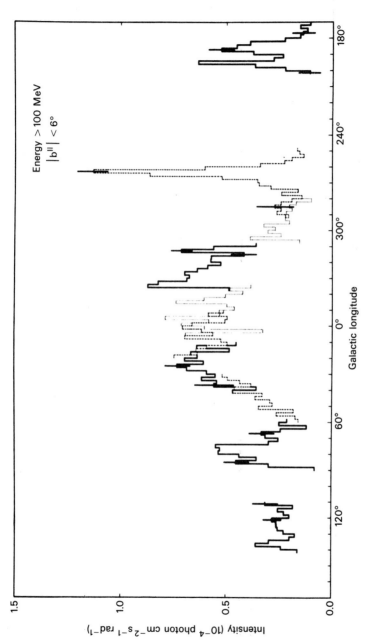

FIG. 9.9. COS-B measurements of the distribution of gamma ray emission ($\varepsilon > 100$ MeV) in Galactic longitude integrated over the latitude range $-6° \leqslant b^{II} \leqslant 6°$. (From Bennett *et al.* 1977.)

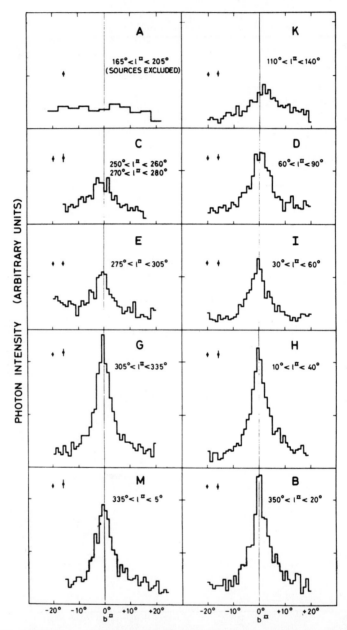

FIG. 9.10. COS-B measurements of the distributions of gamma ray emission ($\varepsilon > 100$ MeV) in Galactic latitude made at 10 directions along the Galactic plane. (From Bennett *et al.* 1977.)

9.4.4. *Gamma ray measurements made in the SAS-2 and COS-B experiments*

Both the SAS-2 and the COS-B experiments detected a strong gamma ray emission from the Galactic disc; a weaker flux, with a steeper energy spectrum, was detected from the whole sky and was assumed to be extragalactic background radiation. The distribution in Galactic longitude of the radiation detected by the SAS-2 telescope at Galactic latitudes between b^{II} = $-10°$ and $b^{II} = +10°$ is shown in Fig. 9.7. Gamma ray emission was detected from the entire disc but there was considerable variation in the intensity with Galactic longitude. Some of the variation was due to discrete sources, and these will be discussed in the next chapter, but there was also an enhanced intensity from $l^{II} \sim 325°$ to $l^{II} \sim 40°$ which was not resolved into individual sources. The width of the emission region in Galactic latitude is shown in Fig. 9.8 for directions towards, and away from, the Galactic centre.

The COS-B measurement of the distribution in Galactic longitude is shown in Fig. 9.9. The distribution is similar to that derived from the SAS-2 measurements and shows the enhancement in directions towards the Galactic centre. Discrete sources are evident at $l^{II} \sim 260°$ and $l^{II} \sim 85°$ whilst the feature at $l \sim 190°$ has been clearly split into two sources by the somewhat better angular resolution of the COS-B telescope. The COS-B latitude distributions, measured at different values of longitude, are shown in Fig. 9.10. Each distribution has been fitted to a gaussian and the full widths at half maximum of the distributions are plotted against Galactic longitude in Fig. 9.11. The narrow distribution obtained from measurements towards the Galactic centre was examined more carefully by considering only gamma rays with energies greater than 200 MeV and the width was shown to be less than 2°.

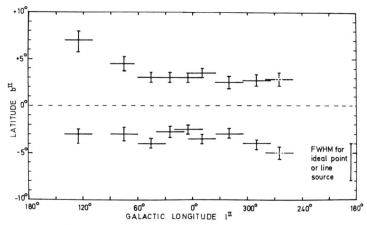

FIG. 9.11. Full widths at half maximum of the Galactic latitude gamma ray distributions as a function of Galactic longitude. (From Bennett *et al.* 1977.)

The energy calorimeter on the COS-B telescope allowed energy measurements to be made up to ∼2000 MeV and the energy spectrum of the radiation from the Galactic disc is shown in Fig. 9.12. The broken line shows the spectrum expected from pion decay and it is clear that the observed spectrum does not have the symmetry about 70 MeV which this mechanism, operating alone, would produce.

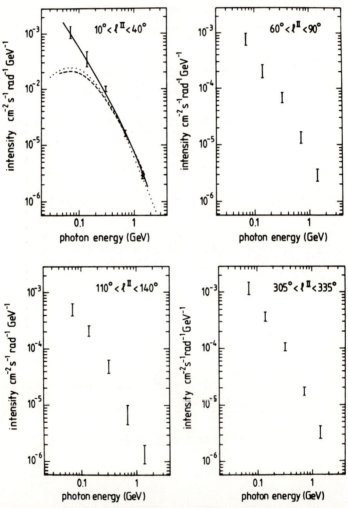

FIG. 9.12. Energy spectra of gamma rays from different directions along the Galactic disc. The dotted and broken lines are spectra predicted by two different models for gamma rays from pion decay whilst the full line is the spectrum predicted when bremsstrahlung is included. (From Bennett *et al.* 1977.)

9.5. Interpretation of the gamma ray measurements from the Galactic disc

Paul, Cassé, and Cesarsky (1976) have compared the latitude and the longitude distributions of the gamma ray emission with those for the radio emission at 150 MHz, and their results are shown in Figs. 9.13 and 9.14. The general agreement between the distributions is good and suggests that the gamma rays, like the radio emission, are produced in cosmic ray interactions in the interstellar medium.

The energy spectrum of the gamma rays provides us with evidence as to which production mechanisms are responsible for the radiation. Stecker (1977) has considered the relative importance in the interstellar medium of three mechanisms—the production and decay of neutral pions, the bremsstrahlung of cosmic ray electrons, and the Compton scattering of electrons on starlight and the 3 K background radiation. The results of these calculations are given in Fig. 9.15 which shows that bremsstrahlung dominates at energies below ~ 70 MeV whereas pion decay dominates at higher energies. Both of these mechanisms will produce a narrow distribution in Galactic latitude because they occur only in regions within ~ 100 pc of the Galactic plane, where the interstellar gas is found. In contrast, the latitude distribution of the gamma rays from Compton scattering is wider because the extent of the region where this process operates is limited only by the

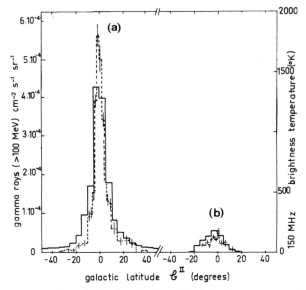

FIG. 9.13. Comparison of the distributions in Galactic latitude of gamma ray emission (broken lines) and 150 MHz radio emission (full lines): (a) Galactic longitude interval $l^{II} = 330°$ to $l^{II} = 30°$; (b) Galactic longitude interval $90° < l^{II} < 170°$ and $200° < l^{II} < 260°$. (From Paul *et al.* 1976, p. 64.) © (1976) The American Astronomical Society.

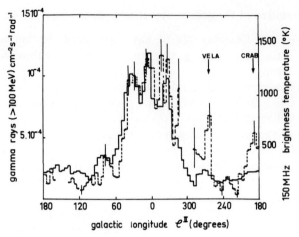

FIG. 9.14. Comparison of the distributions of gamma ray emission (broken line) and 150 MHz radio emission (full line) along the Galactic disc. (From Paul *et al.* 1976, p. 65.) © (1976) The American Astronomical Society.

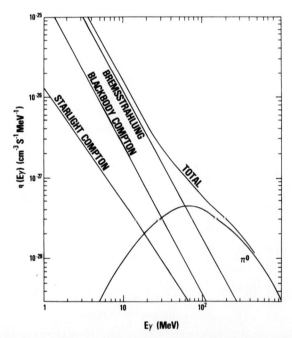

FIG. 9.15. Calculated spectra of gamma rays produced by different mechanisms in the interstellar medium. (From Stecker 1977, p. 61.) © (1977) The American Astronomical Society.

diffusion of cosmic ray electrons; from studies of the radio emission from the Galaxy, Ilovaisky and Lequex (1972) have shown that electrons exist at distances up to ~ 1000 pc from the Galactic plane. The narrowness of the distributions in Fig. 9.10 confirm the conclusion drawn from the calculations of Stecker that Compton scattering does not, in general, make a significant contribution. One exception may be in regions near the Galactic centre where Stecker has shown that part of the enhancement observed in the longitude distribution in this direction may be due to Compton scattering.

All attempts to explain the observed longitude profile of gamma rays in terms of bremsstrahlung and pion decay from a *uniform* distribution of cosmic rays in the Galactic disc have been unsuccessful. Stecker and Jones (1977) have extended this model to include the effects of diffusion of cosmic rays away from their sources. The sources were assumed to be either supernova remnants or pulsars, both of which will have a distribution in the Galaxy similar to that shown in Fig. 9.16; the lack of sources near the centre of the Galaxy is due to the preponderance there of Population II rather than Population I stars. Stecker and Jones use the standard diffusion model to calculate the cosmic ray flux throughout the Galaxy, and from this they calculate the longitude profile of the gamma ray emission. The result of their calculation is shown in Fig. 9.17 and the agreement with the observations is good, apart from a few localized regions of emission which are assumed to be discrete sources. Figure 9.18 shows that an even better agreement is obtained

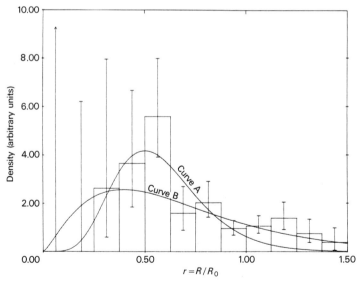

FIG. 9.16. Distribution of pulsars as a function of distance from the centre of the Galaxy. The two curves are best fits to the experimental data, curve B being weighted according to the error bars. (From Stecker and Jones 1977, p. 847.) © (1977) The American Astronomical Society.

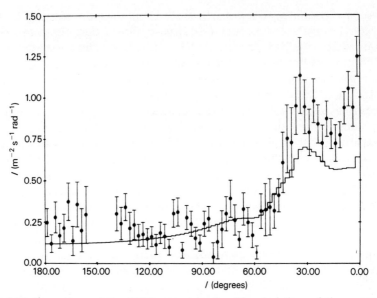

FIG. 9.17. Comparison of the SAS-2 measurements with calculations of the gamma ray emission assuming cosmic ray diffusion from sources distributed according to curve B in Fig. 9.16. The calculated emission is too small in the direction of the Galactic centre. (From Stecker and Jones 1977, p. 854.) © (1977) The American Astronomical Society.

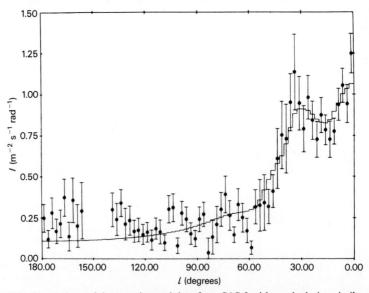

FIG. 9.18. Comparison of the experimental data from SAS-2 with a calculation similar to that used for Fig. 9.17 but with a contribution from Compton scattering which significantly increases the flux from regions near the Galactic centre. (From Stecker and Jones 1977, p. 854.) © (1977) The American Astronomical Society.

if the contribution from Compton scattering near the centre of the Galaxy is included.

Kniffen, Fichtel, and Thompson (1977) have adopted a different approach to the problem of determining the distribution of cosmic rays in the Galaxy. They assume that the çosmic ray flux at any point is proportional to the gas density, and justify this by arguing that the cosmic rays are constrained by magnetic fields which, in turn, are linked to the interstellar gas. In this situation the upper limit to the energy density in the cosmic rays is set by the gravitational field, and hence the density, of the gas. Any region of the Galaxy will continue to accumulate cosmic rays until this limit is reached and the result will be a proportionality between the cosmic ray flux and the gas density. Kniffen *et al.* point out that this condition is known to exist near the solar system and it is therefore reasonable to assume that it occurs throughout the Galaxy. This model predicts that the gamma ray intensity in any direction will be proportional to the square of the column density of the interstellar gas in that direction. Figure 9.19 shows that there is good agreement between this prediction and the observations. It is inevitable that the two different models arrive at very similar results for the distribution of gamma ray emission because both assume that the cosmic ray flux is enhanced in the same regions of the Galaxy, namely those regions which contain gas clouds and Population I stars.

The limited angular resolution of both the SAS-2 and the COS-B telescopes may lead to confusion between discrete sources if the density of sources is high and it is possible that the enhanced intensity of gamma rays

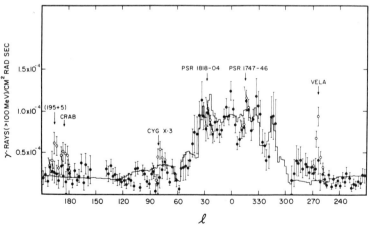

FIG. 9.19. Comparison of the SAS-2 measurements with calculations which assume that the cosmic ray flux in the Galaxy varies in proportion to the density of the interstellar gas. (From Kniffen *et al.* 1977, p. 772.) © (1977) The American Astronomical Society.

from regions near the Galactic centre is really produced by a large number of unresolved sources. This interpretation of the observations will be discussed in the next chapter.

References

Anand, K. C., Daniel, R. R., and Stephens, S. A. (1968). *Nature (London)* **217**, 25.

Bennett, K., Bignami, G. F., Buccheri, R., Hermsen, W., Kanbach, G., Lebrun, F., Mayer-Hasselwander, H. A., Paul, J. A., Piccinotti, G., Scarsi, L., Soroka, F., Swanenburg, B. N., and Willis, R. D. (1977). In *Recent advances in gamma ray astronomy*, ESA SP-124, p. 83. ESA Scientific and Technical Information Branch, Noordwijk.

Browning, R., Ramsden, D., and Wright, P. J. (1971). *Nature (London)* **232**, 99.

Fichtel, C. E., Hartman, R. C., Kniffen, D. A., Thompson, D. J., Bignami, G. F., Ogelman, H., Ozel, M. E., and Tumer, T. (1975). *Astrophys. J.* **198**, 163.

Frye, G. M., Albats, P. A., Zych, A. D., Staib, J. A., Hopper, V. D., Rawlinson, W. R., and Thomas, J. A. (1971). *Nature (London)* **231**, 372.

Hartman, R. C., Kniffen, D. A., Thompson, D. J., Fichtel, C. E., Ogelman, H. B., Tumer, T., and Ozel, M. E. (1979). *Astrophys. J.* **230**, 597.

Ilovaisky, A. and Lequex, J. (1972). *Astron. Astrophys.* **20**, 347.

Kniffen, D. A., Fichtel, C. E., and Thompson, D. J. (1977). *Astrophys. J.* **215**, 765.

Kraushaar, W. L., Clark, G. W., Garmire, G. P., Borken, R., Higbie, P., Leong, V., and Thorsos, T. (1972). *Astrophys. J.* **177**, 341.

Landecker, T. L. and Wielebinski, R. (1970). *Aust. J. Phys. Astrophys. Suppl.* **16**.

Morrison, P. (1958). *Nuovo Cimento* **7**, 858.

Oort, J. H., Kerr, F. J., and Westerhout, G. (1958). *Mon. Not. R. astron. Soc.* **118**, 382.

Paul, J., Cassé, M., and Cesarsky, C. J. (1976). *Astrophys. J.* **207**, 62.

Savage, B. D., Bohlin, R. C., Drake, J. F., and Budich, W. (1977). *Astrophys. J.* **216**, 291.

Scoville, N. Z. and Solomon, P. M. (1975). *Astrophys. J.* **199**, L105.

Shapiro, M. M. and Silberberg, R. (1970). *Annu. Rev. nucl. Sci.* **20**, 323.

Stark, A. A. and Blitz, L. (1978). *Astrophys. J.* **225**, L15.

Stecker, F. W. (1977). *Astrophys. J.* **212**, 60.

—— and Jones, F. C. (1977). *Astrophys. J.* **217**, 843.

Webber, W. R. and Lezniak, J. A. (1974). *Astrophys. space Sci.* **30**, 361.

10

DISCRETE GAMMA RAY SOURCES IN THE GALAXY

10.1. Introduction

In addition to the diffuse radiation detected from the Galactic disc there is also evidence for at least three classes of discrete sources in the Galaxy. The first of these are the sources of high energy gamma rays detected by the SAS-2 and the COS-B telescopes; Swanenburg *et al.* (1981) have shown that the COS-B data contain evidence for 25 statistically significant regions of enhanced emission, the strongest of these being associated with pulsars.

Low energy gamma ray telescopes have detected emission which includes the spectral line at 0.51 MeV from the direction of the Galactic centre. Because of the poor angular resolution of the telescopes it is not yet possible to determine whether this is radiation from one or more discrete sources or whether it is diffuse emission from the interstellar medium.

A third category of source is that which produces short bursts of low energy gamma rays, the duration of a burst sometimes being as short as a few seconds. These sources have not yet been identified but the short timescale of a typical burst limits the dimensions of the source to that of a large star and it is therefore probable that the sources lie in the Galaxy.

10.2. Pulsars and supernova remnants

10.2.1. *General characteristics of pulsars*

A survey of the observational data on pulsars and their theoretical interpretation has been given by Radhakrishnan (1982). Some 320 pulsars have been detected in the Galaxy by radioastronomers but there are reasons to believe that many more remain undetected and that the total number may be as high as a hundred thousand.

The characteristic signal from a pulsar is a series of regularly spaced pulses with a repetition period of the order of a second or even less. The pulses are very short and typically occupy only ~ 3 per cent of the cycle. The intense theoretical activity which followed the discovery of the first pulsars in 1968 quickly identified neutron stars as the most probable sources of the radiation and the pulse profiles favoured models in which the signal is produced by an off-axis beam of radiation emerging from a rapidly rotating star. This interpretation was supported by the subsequent discovery that the periods of all pulsars are increasing, which could be understood as the result of the loss of rotational energy by the star. The two pulsars with the shortest periods,* PSR 0531 + 21 and PSR 0833-45, are therefore considered to be the youngest pulsars; it is presumably not a coincidence that these are the only pulsars which have been detected outside the radio region of the spectrum.

It is generally accepted that neutron stars are formed in the ultimate gravitational collapse of massive stars, events which are also thought to produce supernova explosions. Thus, long before the discovery of pulsars it was assumed that neutron stars would be found in supernova remnants, but only two pulsars, in fact, are associated with remnants. Nevertheless it is considered probable that many, if not all, supernova remnants contain rapidly rotating neutron stars and that these stars play an important role in energetics of the remnants.

Supernova remnants are strong sources of synchrotron radiation which implies that they contain large fluxes of relativistic electrons and, presumably, protons and nuclei. Interactions of cosmic ray particles are known to result in significant fluxes of gamma rays from the Galactic disc, and supernova remnants are therefore expected to be strong discrete sources of gamma ray emission.

10.2.2. *The pulsar PSR0531 + 21 and the Crab nebula*

The angular diameter of the Crab nebula is of the order of an arc minute which is too small to be resolved by present-day gamma ray telescopes. Consequently the emission from the nebula can be distinguished from that of

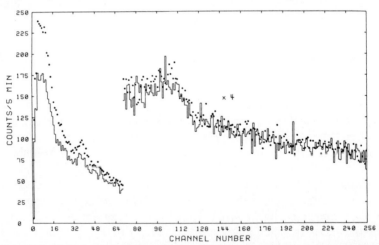

FIG. 10.1. Counting rates recorded from the Crab nebula at energies between 0.056 MeV and 0.96 MeV. Filled circles represent the combined signal from source and background whilst the histogram represents background alone. (From Walraven *et al.* 1975, p. 506.) © (1975) The American Astronomical Society.

* The millisecond pulsars, the first of which was discovered in 1983, seem to be a distinct class and their gamma ray characteristics may be very different from those of normal pulsars.

the pulsar PSR0531 + 21, which lies in the nebula, only by assuming that all the radiation from PSR0531 + 21 is pulsed at its characteristic frequency. The time structure of the radiation from PSR0531 + 21 also allows it to be distinguished from local background radiation and this feature was used in the first measurement of the low energy gamma ray flux from the pulsar which was made with a large unshielded detector with virtually no angular resolution (Hillier, Jackson, Murray, Redfern, and Sale 1970).

Later measurements have used scintillation counters inside massive anticoincidence shields; one such telescope, constructed by a group at Rice University in Texas, was described in Section 4.5. This telescope was used to make measurements on the Crab nebula during a balloon flight in May 1973 at an atmospheric depth of 4 g cm^{-2}. The observing programme, which lasted 3 h, consisted of a series of 10 min intervals in which the telescope was pointed alternatively towards the source and then towards a direction 180° away in azimuth to measure the background intensity. The spectra recorded from the two directions, in two different energy intervals, are compared in Figs. 10.1 and 10.2. These results show that, even with a heavily shielded detector, the signal from the source is small compared with the background. The spectrum of the total radiation detected from the Crab (i.e. both pulsar and nebula) is shown in Fig. 10.3, along with data from other experiments. At energies between 0.1 MeV and 1.0 MeV the spectrum can be represented by

$$n(\varepsilon) = 12\varepsilon^{-2.16} \text{ photons cm}^{-2}\text{ s}^{-1}\text{ keV}^{-1},$$

where ε is measured in keV.

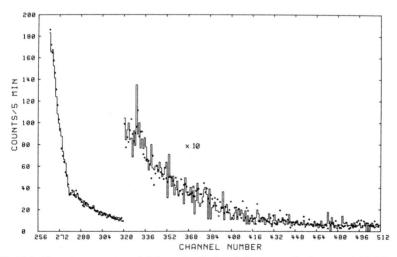

FIG. 10.2. Counting rates recorded from the Crab nebula at energies between 1.0 MeV and 12.1 MeV. The full circles represent the combined signal from source and background whilst the histogram represents background alone. (From Walraven et al. 1975, p. 507.) © (1975) The American Astronomical Society.

The component of the radiation from PSR0531 + 21 was determined by overlaying the data at the characteristic frequency of the pulsar and Fig. 10.4 shows the pulse profile which was obtained. This analysis used data covering the entire energy range from 0.056 MeV to 12.11 MeV but the shape of the profile is essentially determined by photons with energies below a few hundred keV because the intensity falls rapidly with increasing energy. The spectrum of the pulsed radiation is shown in Fig. 10.5; no significant emission was detected above 0.4 MeV and below this energy the spectrum could be represented by

$$n(\varepsilon) = 2.2\varepsilon^{-2.2} \text{ photons cm}^{-2} \text{ s}^{-1} \text{ keV}^{-1},$$

where ε is measured in keV.

High energy gamma rays were detected from the direction of the Crab nebula by the SAS-2 and COS-B telescopes. Figure 10.6 is a map of the sky near the direction of the Galactic anticentre, drawn from data recorded by the COS-B telescope, and it shows evidence for two strong sources, one coinciding with the Crab nebula at $l^{II} = 185$, $b^{II} = -6°$, and the other being an unidentified source at $l^{II} = 195°$, $b^{II} = +5°$. When the data from the direction of the Crab nebula were overlayed at the frequency of PSR0531 + 21 the pulse profile shown in Fig. 10.7 was obtained. This indicates that ~ 15 per cent of the high energy gamma ray flux is unpulsed but it is not

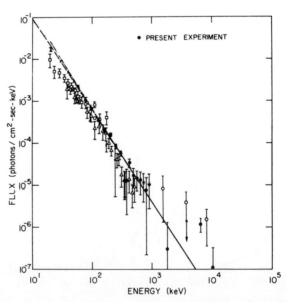

FIG. 10.3. Spectrum of the total radiation, pulsed and unpulsed, from the Crab nebula. The line corresponds to the spectrum $\eta(\varepsilon) = 12\varepsilon^{-2.16}$ where ε is measured in keV. (From Walraven et al. 1975, p. 508.) © (1975) The American Astronomical Society.

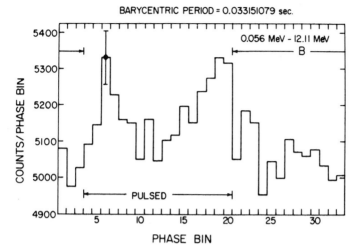

FIG. 10.4. The pulse profile obtained by overlaying the data recorded from the Crab nebula at the repetition rate of the pulsar PSR 0531 + 21. The data corresponds to photons in the energy interval from 0.056 MeV to 12.1 MeV. (From Walraven *et al.* 1975, p. 508.) © (1975) The American Astronomical Society.

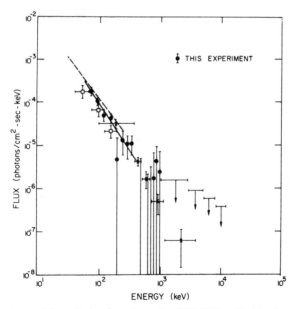

FIG. 10.5. Spectrum of the radiation from the pulsar PSR 0531 + 21. The line corresponds to the spectrum $n(\varepsilon) = 2.2\varepsilon^{-2.2}$, where ε is measured in keV. (From Walraven *et al.* 1975, p. 509.) © (1975) The American Astronomical Society.

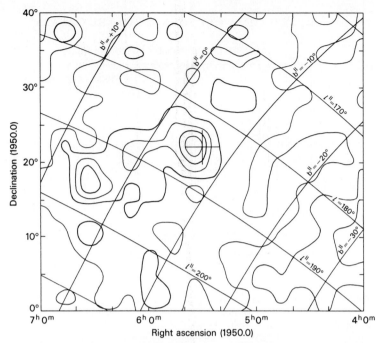

FIG. 10.6. Map drawn from the COS-B measurements of the gamma ray intensity from the direction of the Galactic anticentre. Two localized sources are evident, the Crab at $\alpha \sim 6\,h$ 30 min. (From Bennett *et al.* 1977a, p. 469.)

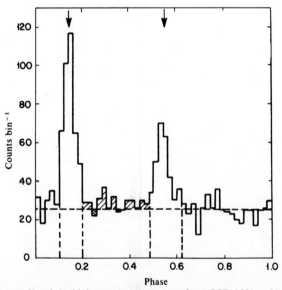

FIG. 10.7. Pulse profile of the high energy gamma rays from PSR 0531 + 21 recorded in the COS-B experiment. The broken horizontal line shows the intensity of the background radiation and the shaded area indicates an unpulsed component of the radiation. (From Wills *et al.* 1982, p. 724.)

known whether this component comes from the pulsar or from the nebula. Five separate observations were made with the COS-B telescope (in August 1975, October 1976, February 1979, September 1979, and September 1980) and there is evidence for a change in the shape of the pulse profile between August 1975 and October 1976 with the secondary pulse stronger in the first observation than in the subsequent ones (Wills *et al.* 1982).

Gamma rays with energies greater than 240 MeV were detected from PSR0531 + 21 by a telescope flown on a balloon in October 1971 (McBreen, Ball, Campbell, Greisen and Koch 1973). The design of the telescope is shown in Fig. 10.8. Primary gamma rays passed through an anticoincidence counter and converted into electron–positron pairs in a lead sheet which was $\frac{1}{16}$ in thick and 4.4 m^2 in area. The electrons then entered a cylinder of Freon gas at a pressure of 1.5 lbf in^{-2}. Cerenkov light from the electrons was collected by a mirror at the far end of the cylinder and focused onto an aperture on the photocathode of a photomultiplier tube. The aperture accepted Cerenkov light from electrons travelling within an angle of 0.85° from the telescope axis. The threshold energy for a particle to produce Cerenkov light in the Freon gas was $\sim 70\ Mc^2$, where M is the mass of the particle. This threshold greatly reduced the background counting rate from protons and mesons which is a major problem with many forms of counter telescope.

On the first flight measurements were made for a period of 3 h 30 min at an atmospheric depth of 7.3 g cm^{-2}. The pulse profile obtained when the data

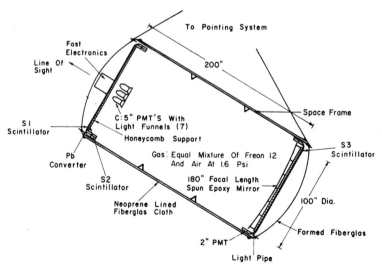

FIG. 10.8. Diagram of a gamma ray telescope designed to detect photons with energies greater than 240 MeV. (From Albats *et al.* 1971, p. 190.)

was analysed at the frequency of PSR0531 + 21 is shown in Fig. 10.9. The positions of the main and secondary pulses, predicted from optical observations, are shown and there is clear evidence of a peak in the gamma ray intensity at each position. The probability that these peaks are merely fluctuations in a flat distribution is less than 1 in 10^6.

The telescope was flown again in July 1973. On the second flight both the observing time and the float altitude were greater but the pulse profile showed no significant peaks and it appears that the gamma ray flux had decreased by a factor of ~2 above 240 MeV and a factor of ~7 above 800 MeV in the interval between the two flights.

Time variations may also account for the results of measurements of the flux of very high energy gamma rays from PSR0531 + 21 which, in some cases, seem to be contradictory. For example, Grindlay, Helmken, and Weekes (1976) have published measurements made at Mount Hopkins, Arizona, with ground-based Cerenkov counters which have a threshold energy of ~8×10^5 MeV. Data recorded in December 1973 showed evidence of a single peak in the pulse profile, the peak occurring 6.5 ms after the predicted position of the main pulse. The probability that the peak was caused by a Poissonian fluctuation was less than 1 in 5000. The peak was not present in data recorded in November–December 1971 nor in February 1973.

The spectrum of PSR0531 + 21 at energies between 0.1 MeV and 10^7 MeV is shown in Fig. 10.10, the data coming from the measurements described in

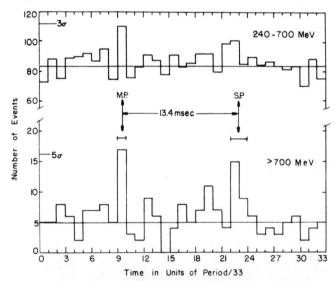

FIG. 10.9. Pulse profile of high energy gamma rays detected from PSR 0531 + 21 by the telescope shown in Fig. 10.8. The absolute phases of the main and secondary pulses were predicted from optical observations of the pulsar. (From McBreen *et al.* 1973, p. 573.) © (1973) The American Astronomical Society.

this section and from similar experiments. The broken line represents the spectrum

$$n(\varepsilon) = 1.0\varepsilon^{-2.1} \text{ photons cm}^{-2}\,\text{s}^{-1}\,\text{keV}^{-1},$$

where ε is measured in keV. The interpretation of this spectrum, together with those of other pulsars, will be considered in the next section.

At gamma ray energies the intensity of the unpulsed radiation from the nebula is less than that of the pulsed radiation from the pulsar and is correspondingly more difficult to detect, but even the upper limits to the flux at higher energies can be used to place restrictions on the conditions in the nebula. There is considerable evidence that radiation throughout the electromagnetic spectrum is produced by relativistic electrons moving in the magnetic field of the nebula. The evidence is strongest at radio and optical frequencies where there is no adequate explanation in terms of thermal emission and where measurements of the polarization allows the magnetic field in the nebula to be mapped. The polarization of the integrated X-ray emission from the nebula has been measured (Weisskopf, Silver, Kestenbaum, Long, and Novick 1978) and it agrees with that expected from the optical measurements. No measurements of the gamma ray polarization have been made but the way in which the low energy gamma ray measurements lie on the extrapolation of the X-ray spectrum suggests that a single mechanism is responsible for the entire spectrum.

Greisen (1971) has reviewed the evidence we have for the strength of the magnetic field in the nebula. An upper limit is set by the polarization measured at optical and radio frequencies. The measurements integrate the polarization of radiation emitted at all depths in the nebula and the absence

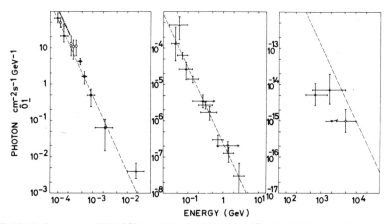

FIG. 10.10. Spectrum of PSR 0531 + 21 from 0.1 McV to 10^7 MeV. The broken line represents the spectrum $\eta(\varepsilon) = 1.0\varepsilon^{-2.1}$, where ε is measured in keV. (From Bennett *et al.* 1977*b*, p. 284.)

of significant depolarization of the escaping radiation by the Faraday effect indicates that the magnetic field strength is less than $\sim 3 \times 10^{-4}$ G. The gamma ray measurements provide a lower limit to the field strength. Gould (1965) has shown that the relativistic electrons which produce the optical synchrotron radiation will interact again with a fraction of the photons through Compton scattering, producing high energy gamma rays. If the magnetic field in the nebula is weak then the fluxes of relativistic electrons must be correspondingly large to produce the observed synchrotron radiation and these electrons will, in turn, produce large fluxes of high energy gamma rays. Thus the flux of high energy gamma rays is inversely correlated with the strength of the magnetic field in the nebula, and the upper limits set to the flux of high energy gamma rays indicate that the magnetic field must be greater than $\sim 3 \times 10^{-5}$ G.

Synchrotron radiation represents a major drain on the energies of the relativistic electrons. The characteristic time, τ_s, for an electron to lose a significant fraction of its energy as synchrotron radiation is given by

$$\tau_s \sim \frac{E}{P_s(E, H)},$$

where $P_s(E, H)$ is the power radiated.

We saw in Section 2 that $P_s \propto E^2 H^2$ so $\tau_s \propto E^{-1} H^{-2}$. The electrons in the Crab nebula which radiate ~ 1 MeV photons have an energy of $\sim 5 \times 10^8$ MeV and, for them, $\tau_s \sim 3$ year. Since this is very much less than the age of the nebula it implies that the acceleration of the electrons could not have occurred in the supernova explosion but must be taking place continuously in the nebula. The discovery of the pulsar in the nebula and its interpretation as a rotating neutron star seemed to provide a site for the acceleration. Gold (1969) pointed out that the slowing down of the pulsar represented a loss of rotational energy by the neutron star of $\sim 10^{38}$ erg s^{-1} which is of the same order as the total luminosity of the nebula. This coincidence could be explained if the slowing down of the pulsar was caused by the acceleration of electrons which then lost their energy as synchrotron radiation in the nebula. In this model the distances which the electrons can travel from the pulsar will be limited by their energy losses, the more energetic electrons reaching only smaller distances because of their shorter lifetimes. Since the more energetic electrons radiate higher energy photons this should result in the size of the radiation source decreasing with photon energy. Wilson (1972) has shown that the sizes of the radio, optical, and X-ray sources can be explained by this model only if it is assumed that the diffusion of the electrons is energy dependent; there is, as yet, no measurement of the angular size of the gamma ray source.

We saw in Chapter 2 that young supernova remnants are possible sources of gamma ray spectral lines if, as is generally accepted, the synthesis of very heavy elements takes place in a star just prior to a supernova explosion. The most intense gamma ray lines will be those from radioactive nuclides with lifetimes of the same order as the age of the remnant and Clayton and Craddock (1965) have calculated the fluxes of these lines expected at the earth from the Crab nebula. There have been several attempts to detect these lines using both scintillation counters and solid state detectors but, as is shown in Table 10.1, these experiments have produced only upper limits which are still orders of magnitude above the predicted intensities and do not place any restraints on models of supernova explosions.

Table 10.1. Upper limits to the intensities of gamma ray lines received from the Crab nebula

Photon energy (MeV)	Source	Fluxes at Earth ($photons\ cm^{-2}\ s^{-1}$)		
		Predicted (Clayton and Craddock 1965)	Observed upper limits	
			Walraven et al. 1975	Levental et al. 1977
0.18	^{251}Cf	1.9×10^{-5}	1.3×10^{-3}	1.5×10^{-3}
0.39	^{249}Cf	1.0×10^{-4}	1.0×10^{-3}	1.5×10^{-3}
0.51	e^+-e^-	—	1.1×10^{-3}	2.4×10^{-3}
0.61	^{214}Bi	7.4×10^{-6}	1.4×10^{-3}	1.3×10^{-3}
1.12	^{214}Bi	3.2×10^{-6}	1.6×10^{-3}	1.4×10^{-3}
1.76	^{214}Bi	4.0×10^{-6}	1.4×10^{-3}	1.0×10^{-3}

10.2.3. *The pulsar PSR0833-45*

High energy gamma rays have been detected from the Vela pulsar PSR0833-45 which is the second fastest pulsar in the sky with a period of 89.2 ms. The pulse profile measured with the COS-B telescope is shown in Fig. 10.11, together with the profiles measured in the radio and optical regions of the spectrum. Although the pulsar period is identical in different regions of the spectrum, as would be expected if the period is determined by the rotation rate of a neutron star, the pulse profiles vary considerably and this suggests that different mechanisms are responsible for the generation of the radiation. The pulse profiles at gamma ray energies of pulsars PSR0531 + 21 and PSR0833-45 are compared in Fig. 10.12 and they are remarkably similar. In both cases there are two strong pulses, separated in phase by an interval which is 42 per cent of the period. The COS-B measurements of the energy spectrum of PSR0833-45 are shown in Fig. 10.13 and the data are well fitted by a simple power law spectrum with an index of -1.89. However, unlike PSR0531 + 21 this power law does not extend to lower energies and PSR0833-44 has not been detected in either the low energy gamma ray or the X-ray regions of the spectrum.

It would seem significant that the two pulsars which have been detected at gamma ray energies are those with the shortest periods and which are therefore presumed to be the most recently formed. Higdon and Lingenfelter (1976) have suggested that the enhanced gamma ray flux from the Galactic disc at longitudes between $l^{II} \sim 270°$ and $l^{II} \sim 60°$ may be due to a large number of unresolved young pulsars in the central regions of the Galaxy. They estimated the contribution from these pulsars by means of a simple model of a pulsar in which the gamma ray luminosity, L_γ, is proportional to

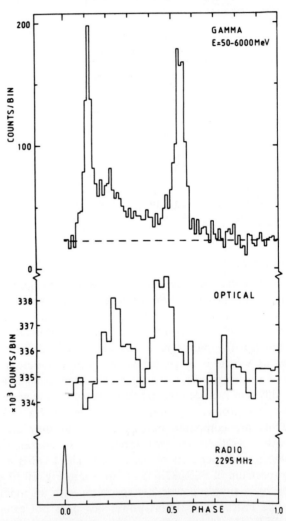

FIG. 10.11. COS-B measurement of the pulse profile of high energy gamma rays from PSR 0833-45 with the optical and radio data for comparison. (From Kanbach *et al.* 1980.)

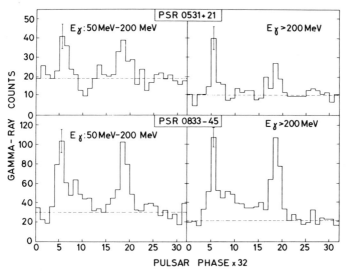

FIG. 10.12. Comparison of the pulse profiles of the high energy gamma rays from PSR 0531 + 21 and PSR 0833-45. (From Bennett *et al.* 1977*b*, p. 281.)

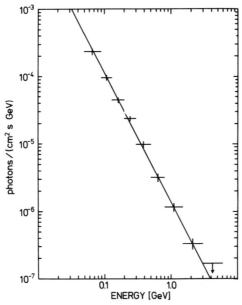

FIG. 10.13. Energy spectrum of the high energy gamma rays from PSR 0833-45. The solid line represents the spectrum $\eta(\varepsilon) = 0.3\varepsilon^{-1.89}$, where ε is measured in keV. (From Kanbach *et al.* 1980.)

the rate of loss of rotational kinetic energy. This gives

$$\frac{\mathrm{d}}{\mathrm{d}t}\left(\tfrac{1}{2}I\omega^2\right) \propto \frac{1}{P^3}\frac{\mathrm{d}P}{\mathrm{d}t},$$

where P is the period of rotation. They used the measurements on the Vela pulsar to derive the constant of proportionality in the relationship. Since neutron stars are probably formed in the final stages of the evolution of massive stars and since the time scale for the evolution of these stars is very short, neutron stars should be formed in regions where normal star formation is still taking place. There is observational evidence that star formation occurs in molecular clouds and we have seen in the previous chapter that there is evidence that the column density of molecular clouds is greatest at just those values of Galactic longitude where gamma ray emission is strongest. By assuming that the birthrate of neutron stars in the Galaxy is one every 30 years Higdon and Lingenfelter calculated the contribution which unresolved pulsars would make to the emission from the Galactic disc. The result is shown in Fig. 10.14, where the broken line is the predicted emission from cosmic ray interactions in the interstellar medium, assuming the cosmic ray flux to be constant throughout the Galaxy, and the dotted line shows the result of including the contribution from pulsars. In this model 90 per cent of the contribution from pulsars comes from young pulsars with periods less than 60 ms. At present long integration times are needed to detect the gamma ray emission from an individual pulsar and this requires a knowledge of its period, which is usually derived from radio observations. Unfortunately the radio signals from pulsars near the Galactic centre are difficult to detect

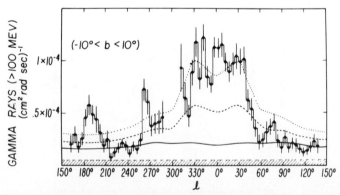

FIG. 10.14. Calculated contribution of pulsars to the gamma ray emission from the Galactic plane. The data points are the COS-B measurements; the full line is the estimated flux from cosmic ray interactions on atomic hydrogen only, the broken line includes similar interactions on molecular hydrogen and the dotted line includes the contribution from unresolved pulsars. (From Higdon and Lingenfelter 1976, p. L108.) © (1976) The American Astronomical Society.

because the pulses are smeared out by the large dispersion in the interstellar gas. The detection of the gamma ray emission from such pulsars will have to wait until the advent of gamma ray telescopes with much greater sensitivity which can detect the signals in a shorter integration time.

10.2.4. *The mechanism responsible for the gamma ray emission from pulsars*

Pulsars were identified as rotating neutron stars soon after their discovery in 1967 but the explanation of the mechanism which produces their radiation has proved much more difficult. One of the most striking features of the pulsar radiation is the way in which the width of the main pulse remains essentially constant throughout the electromagnetic spectrum, in some cases over a frequency range of more than 10^{14} to 1; at the same time the energy spectrum suggests that different mechanisms are operating in different regions of the spectrum and it therefore appears that the pulse profile must be produced by the geometry of the source. Most models take the magnetic field of the neutron star to be the source of the radiation and, to create a pulsed signal at the earth, assume that the star's magnetic axis is misaligned with its rotation axis. The magnetic field at the surface is very strong, perhaps as high as $\sim 10^{12}$ G, and it has an approximate dipole configuration. At greater distances from the star, where the rotational speed approaches the velocity of light, the magnetic field becomes distorted and it can be shown that the field lines which actually pass through the speed-of-light cylinder must be open lines which do not return to the star.

Sturrock (1971) has proposed a mechanism by which particles can be accelerated in the magnetic field around a neutron star and this mechanism has been a starting point for a number of theories of the radiation. The open field lines will be twisted into a spiral and this component of the field, which has a non-zero value of curl **B**, will have electric currents associated with it; these currents will take the form of streams of electrons and protons emerging from the polar caps. The particles will move along the field lines since any motion across the lines will be quickly damped by synchrotron radiation. However, even motion along the field lines leads to radiation because the field lines are curved. The spectrum of this curvature radiation is similar to that of bremsstrahlung and the energies of the photons radiated extend up to the energy of the particle. The high energy photons of the curvature radiation subsequently undergo pair production in interactions not with matter but with the intense magnetic field. The cross-section for pair production in a magnetic field is proportional to $B_{\perp}$, the component of the magnetic field perpendicular to the direction of the photon. At the point of production $B_{\perp}$ is zero since the photons are emitted along the field direction, but the photons travel in straight lines whereas the field lines are curved so that the value of $B_{\perp}$ increases until pair production occurs. The electrons created through pair production in turn produce curvature radiation and an electromagnetic cascade develops. Sturrock claims that it is the radiation from this cascade which forms the gamma ray beam from the pulsar. Salvati and Massaro (1978) have calculated the gamma ray spectrum predicted by this model and have shown that it is in good agreement with the spectrum measured from PSR0833-45. Although the gamma rays from PSR0531-21 may be produced in a similar way, the optical and X-ray emission from this pulsar would require a separate mechanism.

10.3. Molecular clouds in the Galaxy

A discrete source of high energy gamma rays detected by the COS-B experiment at $l^{II} = 353.3°$, $b^{II} = +16.0°$ coincides with the direction of the dark molecular cloud complex near the star ρOph (Mayer-Hasselwander et al. 1980). With the limited angular resolution of the COS-B telescope the identification of the source with the cloud must be regarded as tentative, but at such a high Galactic latitude the number of possible candidates in the Galaxy is small. Bignami and Morfill (1980) have argued that the gamma ray emission comes from cosmic ray interactions with matter in the cloud. The cloud is at a distance of 160 pc from the sun; its mass is uncertain but, taking a probable value of $\sim 3 \times 10^3 M_{\odot}$, the estimated gamma ray emission would be a factor of ~ 5 too small to explain the observations. Bignami and Morfill suggested that the cosmic ray flux may be enhanced within the cloud, an argument which was used in Chapter 9 to explain the distribution in Galactic longitude of the diffuse gamma ray emission from the disc.

Caraveo et al. (1980) have examined the COS-B data for evidence of gamma ray emission from the giant molecular cloud in Orion. This cloud, which is at a distance of 500 pc and has a mass of $\sim 10^5 M_{\odot}$, has an angular diameter of $\sim 10°$ and so should be resolved by the COS-B telescope.

Figure 10.15 shows profiles of the gamma intensity in Right Ascension through the Orion region, drawn at three different values of Declination. These data were used to construct the map of the gamma ray emission from the region which is shown in Fig. 10.16. The emission appears to coincide with the position of the molecular cloud although the statistical weight of the data is low. Carevo et al. calculated that the gamma ray emission could come from cosmic ray interactions in the cloud without invoking an enhancement of the cosmic ray flux.

10.4. Unidentified discrete sources of high energy gamma rays in the Galaxy

The COS-B catalogue (Swanenburg et al. 1981) contained 20 discrete sources which were at Galactic latitudes of less than 5° and which have not been identified. Table 10.2 gives the details of these sources.

An estimate of the distance to a typical source can be obtained from the

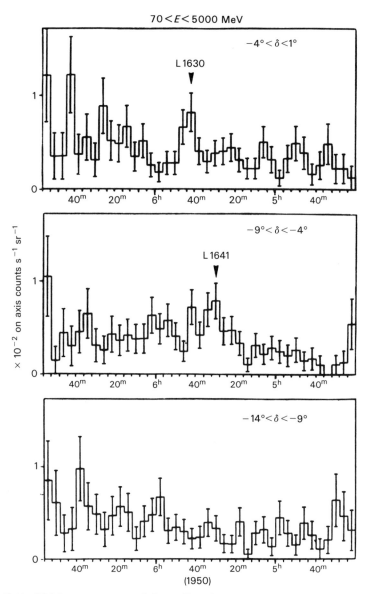

FIG. 10.15. COS-B measurements of the profiles of gamma ray emission across the Orion region. The profiles are drawn in Right Ascension for three values of Declination. (From Caraveo *et al.* 1980, p. L4.)

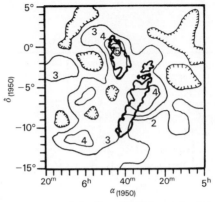

FIG. 10.16. Map of the gamma ray emission from the Orion region drawn from the data in Fig. 10.15. The contour unit is 2×10^{-3} counts s^{-1} sr^{-1}. The heavy lines follow the boundaries of dark clouds. (From Caraveo *et al.* 1980, p. L4.)

Table 10.2. Unidentified sources of high energy gamma rays lying close to the Galactic plane

Source catalogue no.	Position		Flux $> 100\ MeV$ $(10^{-6}\ photons\ cm^{-2}\ s^{-1})$
	l^{II} (deg)	b^{II} (deg)	
2CG 006 − 00	6.7	−0.5	2.4
2CG 013 + 00	13.7	0.6	1.0
2CG 036 + 01	36.5	1.5	1.9
2CG 054 + 01	54.2	1.7	1.3
2CG 065 + 00	65.7	0.0	1.2
2CG 075 + 00	75.0	0.0	1.3
2CG 078 + 01	78.0	1.5	2.5
2CG 095 + 04	95.5	4.2	1.1
2CG 121 + 04	121.0	4.0	1.0
2CG 135 + 01	135.0	1.5	1.0
2CG 195 + 04	195.1	4.5	4.8
2CG 218 − 00	218.5	−0.5	1.0
2CG 235 − 01	235.5	−1.0	1.0
2CG 284 − 00	284.3	−0.5	2.7
2CG 288 − 00	288.3	−0.7	1.6
2CG 311 − 01	311.5	−1.3	2.1
2CG 333 + 01	333.5	1.0	3.8
2CG 342 − 02	342.9	−2.5	2.0
2CG 356 + 00	356.5	0.3	2.6
2CG 359 − 00	359.5	−0.7	1.8

distribution of the sources in Galactic longitude and latitude. The absence of a strong concentration, in longitude, of sources within $\sim 30°$ of the Galactic centre shows that sources as far away as the centre have not been detected and this sets an upper limit to the distance of ~ 7 kpc. On the other hand the narrow distribution, with a standard deviation of $1.1°$, in latitude sets a lower limit to the distance. If the scale height of the sources on either side of the Galactic plane is ~ 40 pc, which is typical of young stellar populations, then the distances to the sources cannot be less than ~ 2 kpc. These limits imply that the gamma ray luminosity, L_y, of a typical source lies in the range 4 $\times 10^{35}$ erg s^{-1} to 5×10^{36} erg s^{-1}. Swanenburg et al. (1981) show that the absence of X-ray or radio counterparts to the sources implies that $L_y > 10 L_x$ and $L_y \gg L_{radio}$. These results impose considerable restrictions on models of the sources.

The geometrical arguments given above suggest that the total number of sources in the Galaxy must be more than ~ 100; on the other hand, the number cannot be greater than ~ 1000 if the integrated emission from the sources is not to exceed the observed emission from the Galactic disc.

10.5. Gamma ray spectral lines from the central region of the Galaxy

The establishment of criteria by which to judge the validity of claims for the detection of gamma ray lines from cosmic sources has proved to be even more difficult than the corresponding task for continuous spectra. The reason for this is that a very large number of lines are possible, although not equally probable, and this allows the data to be searched over a wide energy range for excess counting rates in just a few channels. Cherry, Chupp, Dunphy, Forrest, and Ryan (1980) have shown that this freedom to search for lines means that many of the claimed detections of lines are merely statistical fluctuations. Another problem is the fact that the strongest line expected from many sources is the electron–positron annihilation line at 0.51 MeV; this line also occurs in the locally produced background radiation and so special care must be taken when subtracting background from an observation.

Johnson and Haymes (1973) reported the detection of a spectral line at 472 $\pm$ 24 keV from the direction of the Galactic centre on two balloon flights in November 1970 and November 1971. The detector was an actively shielded scintillation counter, similar to that illustrated in Fig. 4.6. The intensity of the line was $(1.8 \pm 0.5) \times 10^{-3}$ photons cm^{-2} s^{-1}. Using a somewhat larger central detector on a balloon flight in April 1974 the same group measured a line at 530 ± 11 keV with an intensity of $(0.8 \pm 0.2) \times 10^{-3}$ photons cm^{-2} s^{-1} (Haymes et al. 1975). Cherry et al. (1980) pointed out that these features were found after a search around the expected position of the annihilation line at 511 keV, and when the number of possible positions for an excess counting rate is taken into account the results are not statistically significant.

Leventhal, MacCallum, and Stang (1978) flew an actively shielded semiconductor detector on a balloon in November 1977 and observed a line at 511 keV from the Galactic centre region. The design of the telescope, which had a 130 cm³ Ge(Li) crystal as its central detector, is shown in Fig. 10.17. The data recorded in the energy range from 430 keV to 540 keV are shown in Fig. 10.18; the full circles are the counting rates from the direction of the Galactic centre whilst the open squares are the rates recorded from a direction at the same zenith angle but 180° away in azimuth. The lines at 439 keV and 472 keV are merely background radiation, with equal intensities in the two directions, but the line at 511 keV also has a primary component towards the Galactic centre where its intensity is greater. The excess counting rate in the line from the Galactic centre was 4.5 standard deviations and corresponded to an intensity of $(1.2 \pm 0.2) \times 10^{-3}$ photons cm^{-2} s^{-1}. The telescope was reflown in April 1979 and again detected the line at 511 keV, this time with an intensity of $(2.4 \pm 0.7) \times 10^{-3}$ photons cm^{-2} s^{-1}. The width of the line was less than 3.5 keV, the limit set by the energy resolution of the telescope.

In addition to the spectral line, the continuous spectrum shown in Fig. 10.19 was detected from the Galactic centre in these experiments. This spectrum probably contains contributions from one or more of the known X-ray sources in this region of the sky but it may also contain radiation from the

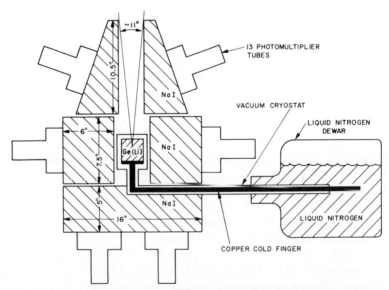

FIG. 10.17. Diagram of the actively shielded Ge(Li) crystal used in measurements of annihilation radiation from the Galactic centre. (From Leventhal, McCallum, and Watts 1977, p. 492.) © (1977) The American Astronomical Society.

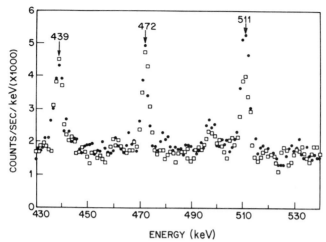

FIG. 10.18. Spectra measured on a balloon flight of November 1977. The full circles represent data recorded from the direction of the Galactic centre and the open squares represent background data. (From Leventhal *et al.* 1978.) © (1978) The American Astronomical Society.

three-photon annihilation process which, as we saw in Section 2.2.5, is important if positronium is formed before annihilation takes place. Leventhal (1973) has shown that positronium formation will be significant if annihilation takes place in a medium in which the density is less than $\sim 10^{15}$ atoms cm^{-3} and the temperature is less than $\sim 10^6$ K.

If it is assumed that the source of the annihilation radiation lies near the Galactic centre, the measured flux indicates that the annihilation rate is $\sim 2 \times 10^{43}$ positrons s^{-1}. The fact that the source lies in the direction of the Galactic centre may be significant and, for example, may indicate that it is related to the powerful non-thermal sources seen at the nuclei of some external galaxies. On the other hand it may merely reflect the fact that almost all the normal components of the Galaxy have a greater concentration in that direction in the sky. There is some evidence that the intensity of the radiation is variable over a time scale of less than a year (Paciesas *et al.* 1982); if confirmed, this would indicate that the source was less than a light year across and would rule out a source which was distributed throughout the interstellar medium.

One source of positrons in the interstellar medium is the decay of charged pions produced in the nuclear interactions of cosmic ray particles with the interstellar gas. Neutral pions created in the same collisions decay into high energy gamma rays and an upper limit to the strength of this source of positrons can be set by the measurements of high energy gamma rays from the Galactic disc. This limit indicates that less than ~ 2 per cent of the

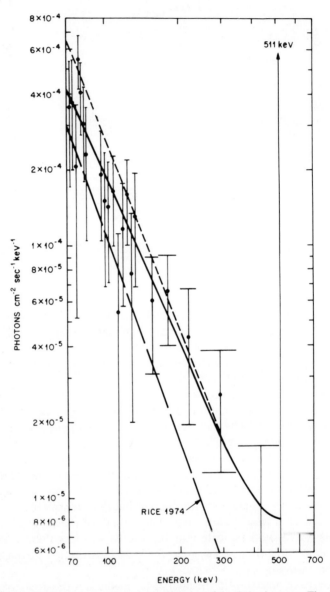

FIG. 10.19. Energy spectra recorded from the direction of the Galactic centre. The data points represent measurements of April 1979 and the solid line has been fitted to these points. The short broken line represents the spectrum measured in November 1977 and the long broken line represents the spectra measured by the Rice group in April 1974. (From Leventhal, MacCallum, Huters, and Strang 1980, p. 341.) © (1980) The American Astronomical Society.

required positrons come from the decay of pions in interstellar space (Ramaty and Lingenfelter 1979).

There is some evidence for a low energy component of the cosmic radiation in interstellar space which does not reach the earth because of the shielding effect of the interplanetary magnetic fields. These particles would have insufficient energy to produce pions in nuclear collisions, but they could raise interstellar nuclei to excited states, some of which may decay with the emission of positrons. Annihilation radiation from this source should be accompanied by other gamma ray lines from nuclear de-excitation, such as the line at 4.44 MeV from ^{12}C, and there is no evidence for the production of these lines. The absence of these lines suggests that less than ~ 30 per cent of the positrons come from nuclei in the interstellar medium which have been excited in collisions with a low energy component of the cosmic rays (Ramaty and Lingenfelter 1979).

There have been many attempts to estimate the radioactivity produced as a result of nucleosynthesis in a supernova explosion. The calculations of the radiation expected are very dependent on the model adopted for a supernova explosion since it is necessary to consider not only the route which the nucleosynthesis takes but also the transparency of the remnant to the radiation produced. There is general agreement that the large abundance of ^{56}Fe in the universe is due to the copious formation of ^{56}Ni in supernova explosions. The ^{56}Ni undergoes two successive β-decays:

$$^{56}Ni + e^- \xrightarrow{\tau = 6.1 \text{ days}} \, ^{56}Co + \nu$$

$$^{56}Co \xrightarrow{\tau = 78.8 \text{ days}} \, ^{56}Fe + e^+ + \nu.$$

If this is, indeed, the main source of ^{56}Fe then the observed abundance of ^{56}Fe implies that the mean production rate of positrons over the lifetime of the Galaxy has been $\sim 2 \times 10^{45}$ positrons s^{-1}. The gamma ray measurements indicate that the observed annihilation rate is $\sim 2 \times 10^{43}$ positrons s^{-1}, implying that on average only ~ 1 per cent of the annihilation radiation need escape from a typical remnant to produce the observed flux. Since the mean lifetime of ^{56}Ni is only 0.31 years and the average interval between supernova explosions in the Galaxy is ~ 30 years, there should be large fluctuations in the intensity of annihilation radiation from sources of this type.

10.6. Sources of gamma ray bursts

In 1973 Klebsedal, Strong, and Olson (1973) reported the discovery of a class of gamma ray sources whose emission characteristics were quite unlike those of any other known astronomical object. The gamma ray luminosity of one of these sources typically rises rapidly from a level which is below the level of detection of present telescopes, reaches a peak, and then fades away after a period which may be as short as a few seconds. Moreover a typical source does not seem to repeat this behaviour, at least over time scales as long as a year. It is interesting to note that a source with this type of behaviour is not easy to detect with conventional optical or radio telescopes because these telescopes monitor only a small fraction of the sky at any time and it is not possible to predict where the burst source will appear. By contrast the field of view of a low energy gamma ray telescope can be made very large without any significant loss in sensitivity because only a small fraction of the background signal is due to radiation entering the acceptance angle of the collimator. We saw in Chapter 4 that a low energy gamma ray telescope can, in some cases, be given a field of view of 4π sr without any loss of sensitivity.

The gamma ray burst sources were discovered with unshielded scintillation counters carried by the Vela series of satellites. These satellites, which were part of a defence programme to monitor nuclear explosions in space, were launched into circular orbits with a radius of $\sim 1.2 \times 10^5$ km and the launch programme was arranged so that there were always at least two working satellites in orbit at any time. A spacecraft carried six caesium iodide scintillation counters, each with a volume of ~ 10 cm^3. Passive shielding ensured that primary electrons with energies below 0.75 MeV and protons below 20 MeV could not penetrate to the detectors. Signals corresponding to energy losses in the detectors between 0.2 and 1.5 MeV were selected for analysis. The counting rate was monitored continuously and whenever there was a significant increase in the rate the subsequent pulses were analysed in a series of time channels whose periods increased logarithmically from 0.1 s to 10 s. These data were stored on the satellite and transmitted to the ground at the end of the event.

Because there were always at least two satellites in operation at any time it was possible to discriminate against instrumental effects, which could masquerade as genuine events, by demanding that a burst should be recorded by at least two satellites. Moreover, since the satellites were, in general, widely spaced it was possible to gain some information on the direction to the source from the difference in the arrival times of the burst at the two satellites; this allowed the cosmic bursts to be distinguished from the most common forms of background events, namely solar X-ray bursts and bursts from the earth's atmosphere caused by precipitation of particles from the radiation belts. If the direction to the source makes an angle θ to the line P_1P_2 joining the two satellites (Fig. 10.20), then the difference in the arrival time of the burst at the satellites is given by

$$\tau = \frac{B \cos \theta}{c}$$

where B is the length P_1P_2. A measurement of τ therefore locates the source to the surface of a cone with a half-angle θ. If data are available from three

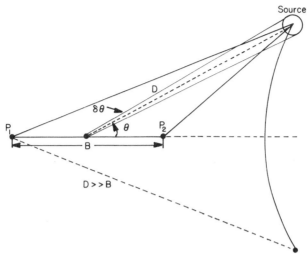

FIG. 10.20. Diagram showing how the direction to a source can be determined from a measurement of the difference in the arrival time of a burst at two separated detectors. (From Chupp 1976, p. 194.)

satellites, two such cones can be constructed and the source must lie in one of the two directions corresponding to the intersections of the cones. The accuracy of this method can be estimated by noting that

$$\Delta \tau = \frac{B \sin \theta}{c} \Delta \theta.$$

With $\Delta \tau \sim 10^{-3}$ s and $B \sim 10^5$ km we find $\Delta \theta \sim 10$ arc min.

A typical gamma ray burst recorded by the Vela satellites is shown in Fig. 10.21. The design of the data analysis did not allow the onset of the burst to be studied. The intensity decreased to a level corresponding to the detection limit of the instrument over a period of ~ 10 s but the fading was not monotonic and several subsidiary bursts were recorded at the times indicated by the arrows in Fig. 10.21. Strong, Klebsedal, and Olson (1974) published a catalogue of the bursts recorded by the Vela satellites from 1967 to 1973. Twenty-four bursts were detected with strengths of 3×10^{-6} erg cm^{-2} or greater. Nine of the bursts were recorded by three satellites and for these it was possible to determine the source direction, apart from the ambiguity mentioned above. These directions seemed to be distributed randomly over the sky and there was little evidence of clustering around the Galactic plane, the values of Galactic latitude ranging from $+39°$ to $-37°$.

Once the Vela detectors had demonstrated the existence of gamma ray bursts it was possible to examine data from other experiments for bursts recorded in coincidence with the Vela detectors. For example, bursts were recorded by the IMP satellites, which contained unshielded scintillation counters designed to study low energy cosmic rays in the solar system, and by the anticoincidence shield of a detector on OSO-7 designed to study X-ray emission from the sun. Cline and Desai (1975) measured the energy spectra of nine bursts, shown in Fig. 10.22, recorded by the IMP-7 satellite. These spectra could all be represented by a common form

$$n(\varepsilon) \sim \exp\left(\frac{\varepsilon}{\varepsilon_0}\right)$$

for $0.1 < \varepsilon < 0.4$ MeV with $\varepsilon_0 \sim 0.150$ MeV, and

$$n(\varepsilon) \sim \varepsilon^{-2.5}$$

for $0.4 < \varepsilon < 1.0$ MeV.

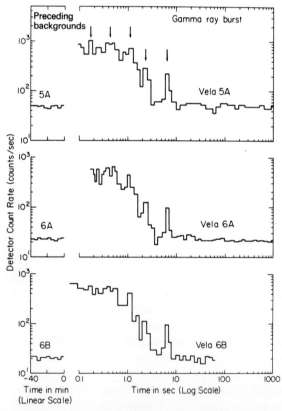

FIG. 10.21. Counting rate as a function of time for a typical gamma ray burst; this burst was recorded by detectors on three Vela satellites on 22nd August 1977. (From Klebsedal *et al.* 1973, p. L87.) © (1973) The American Astronomical Society.

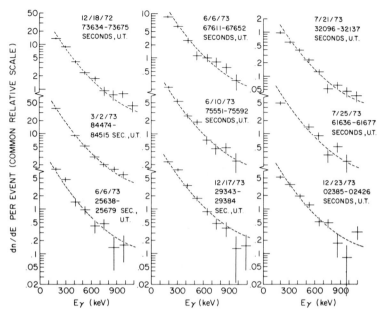

FIG. 10.22. Energy spectra of nine gamma ray bursts recorded by detectors on the IMP-7 satellite. (From Cline and Desai 1975, p. L45.) © (1975) The American Astronomical Society.

Measurements on 85 bursts in 1978 and 1979 were made with the KONUS detectors on the Venera 11 and Venera 12 spacecraft (Mazets *et al.* 1982). The detectors in these experiments were designed so that the data from just one satellite could be used, if necessary, to determine the direction to a source. Each spacecraft contained six sodium iodide detectors, 80 mm in diameter by 30 mm thick. A single detector had passive shielding around the sides and back, and its sensitivity on its front surface varied approximately as $\cos \theta$ where θ was the angle between the axis of the detector and the direction to the source. The six detectors were orientated so that they monitored the whole sky, and by comparing the signals from different detectors it was possible to measure the direction to the source of a strong burst with an accuracy to $\sim 3°$.

The energy spectra of the bursts recorded by the KONUS experiment were of the same form as those measured by Cline and Desai although the value of ε_0 in the exponential region of the spectrum showed some variation from burst to burst. The superior energy resolution of the KONUS detectors revealed narrow features in the continuous spectra which have been interpreted as either spectral lines or absorption dips. Figure 10.23 shows an example of a spectrum with an emission line at $\varepsilon = 45$ keV; Fig. 10.24 shows

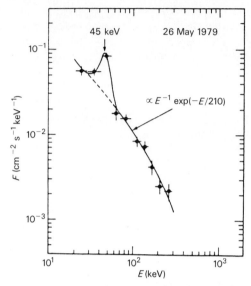

FIG. 10.23. Energy spectrum of a burst recorded on 26th May 1979 by detectors in the KONUS experiment. An emission line is evident at $\varepsilon = 45$ keV. (From Mazets, Golenetskii, Aptekar, Guryan, and Ilyinskii 1981, p. 380.)

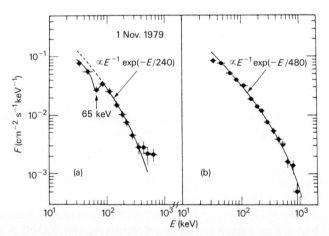

FIG. 10.24. Energy spectra of a burst recorded on 1st November 1979 by detectors on the KONUS experiment: (a) spectrum recorded during the initial 8 s of the burst; (b) spectrum recorded during the subsequent 24 s (note that the absorption feature has disappeared and that the continuous spectrum has hardened). (From Mazets *et al.* 1981, p. 379.)

an absorption feature at $\varepsilon = 65$ keV and also demonstrates, as was often the case, that this feature was present only during the initial phase of the burst.

The true time profile of a burst, not truncated by triggering requirements as in the Vela data, was determined from measurements made on the Apollo 16 spacecraft. The detector was a scintillation crystal 7 cm in diameter × 7 cm thick and data were recorded between 0.067 MeV and 5.1 MeV with high time resolution. The time profile of an event recorded on 27th April 1972 is shown in Fig. 10.25. At least seven distinct peaks can be seen, large variations occurring on time scales of less than a second. This places an upper limit on the dimensions of the source of $\sim 3 \times 10^5$ km.

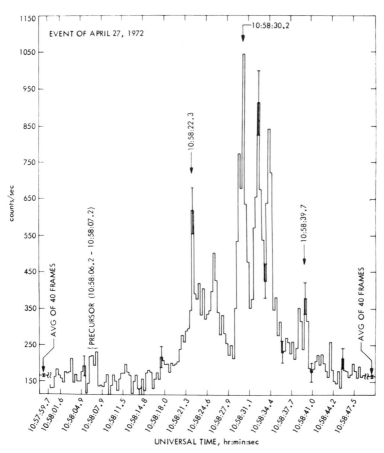

FIG. 10.25. Counting rate as a function of time for a gamma ray burst recorded on 27th April 1972 by a detector on the Apollo 16 spacecraft. Unlike the data shown in Fig. 10.21 the counting rate was recorded during the rise of the burst and there is even evidence of a precursor some 12 s before the main event. (From Metzger, Parker, Gilman, Peterson, and Trombka 1974, p. L21.)
© (1974) The American Astronomical Society.

All attempts to detect optical and radio emission from gamma ray burst sources have been unsuccessful. The failure to detect optical emission may well be due merely to the unpredictability of the bursts and the narrow fields of view of optical telescopes. A search has been made for radio emission at 151 MHz using an aerial system which monitored a large fraction of the sky (Baird *et al.* 1975). No bursts with a strength greater than 10^{-12} ergs cm^{-2} in a 1 MHz bandwidth were recorded. Cavallo and Jelley (1975) have shown that this upper limit to the radio emission excludes stellar flares as the sources of the bursts, if it is assumed that these flares are just scaled-up versions of solar flares.

Some information on the distances to the sources can be gained from a study of the distribution in the strengths observed at the earth. Consider a population of sources, each with an energy output E, which is distributed uniformly in space. The strength, S, of a burst received at the earth from a source at a distance r is given by

$$S = \frac{E}{4\pi r^2}.$$

The number of sources, $N(<r)$, within a distance r will vary as r^3 and all these will have a strength greater than S. So,

$$N(>S) \propto r^3 \propto S^{-3/2}.$$

A similar analysis for sources distributed in a plane gives

$$N(>S) \propto r^2 \propto S^{-1}.$$

Sources distributed uniformly throughout the Galactic disc will appear as a uniform distribution for $r < \beta$, where β is the semi-thickness of the disc, and as a planar distribution for greater distances up to the edge of the Galaxy.

Attempts have been made to extend the observations to bursts with strengths below those detectable on the satellite experiments by using large detectors on high altitude balloons. The rate of bursts recorded by the Vela satellites was ~5 per year and it is necessary to increase this rate by at least two orders of magnitude to observe bursts in a balloon experiment. Despite the use of detectors with areas greater than a square metre most balloon experiments have only been able to set upper limits to the frequency of small bursts but even these lie an order of magnitude below the $S^{-3/2}$ extrapolation of the satellite data (Fig. 10.26). The combined satellite and balloon measurements are best fitted by a model in which the sources are distributed uniformly over a Galactic disc with a semi-thickness of 400 pc (Jennings and White 1980). If the change in slope of the distribution in Fig. 10.26 takes place at a measured strength, S_B, then this should correspond to a source at a distance of ~400 pc. Taking S_B to be ~10^4 ergs cm^{-2}, we see that the total energy in a burst must be ~10^{40} erg. From the observed number of sources

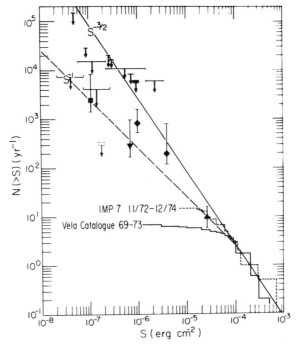

FIG. 10.26. Balloon and satellite measurements of the distribution of apparent burst strengths. (From Jennings and White 1980, p. 111.) © (1980) The American Astronomical Society.

with apparent strength greater than S_B we deduce that the space density of bursts is $\sim 10^{-8}$ pc^{-3} year^{-1}.

The directions of 37 of the bursts detected by the KONUS experiment were measured with an accuracy to $\sim 4°$ and these are plotted in Galactic coordinates in Fig. 10.27. There is no obvious clustering of the sources about the Galactic plane but it should be remembered that these are the stronger bursts, many of which originate from sources at distances less than 400 pc.

A very unusual burst was recorded by a large number of detectors on 5th March 1979 (Cline *et al.* 1980). The burst was one of the most intense so far recorded and it consisted of a short spike, which had a rise time of only 0.25 ms and a total duration of ~ 120 ms, followed by a weaker component which lasted for nearly 3 m. In the weaker component there was clear evidence of oscillations in the signal with a period of 8.00 ± 0.05 s (Fig. 10.28). The sharp spike, together with the fact that the burst was detected by many widely spaced satellites, resulted in a very accurate measurement of the direction to the source; Fig. 10.29 shows that the measured direction

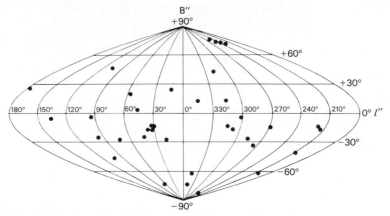

FIG. 10.27. Distribution of the directions, in Galactic coordinates, of 37 bursts recorded by detectors in the KONUS experiment. (From Mazets and Golenetskii 1981, p. 56.)

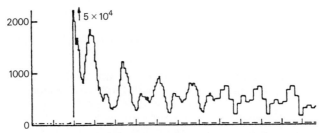

FIG. 10.28. Oscillations with a period of 8 s in the intensity of the unusual burst of 5th March 1979 measured by detectors on the Venera 11 spacecraft. The marks on the horizontal scale are at intervals of 5 s. (From Mazets, Golenetskii, Ilyinskii, Aprekar, and Guryan 1979, p. 588.)

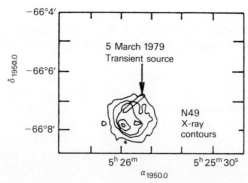

FIG. 10.29. Location of the source of the unusual burst of 5th March 1979 plotted on an X-ray map of the supernova remnant N49 in the Large Magellanic Cloud. (From Cline 1981.)

coincided closely with that of the supernova remnant N49 in the Large Magellanic Cloud.

The energy spectrum of the radiation varied during the burst as is shown in Fig. 10.30. In the initial spike there was evidence for an emission line at $\varepsilon \sim 420$ keV, a feature which has also been observed in some normal bursts; during the oscillations which followed the initial spike the spectrum was softer and the emission line had disappeared. If the source of this burst really did lie in the Large Magellanic Cloud, at a distance of 55 kpc, the total luminosity must have been $\sim 10^{45}$ erg s^{-1}.

Attempts to produce models for the sources of the bursts have been handicapped by the lack of optical identifications. Jennings and White (1980) have pointed out that the space density of the bursts, $\sim 10^{-8}$ pc^{-3} year^{-1}, is such that a source must produce many bursts during its lifetime. For example, even normal stars have a space density of only $\sim 10^{-1}$ pc^{-3} so that such a star would have to produce a burst every $\sim 10^7$ year. Many models invoke neutron stars as the sources of the bursts, and since these stars have a

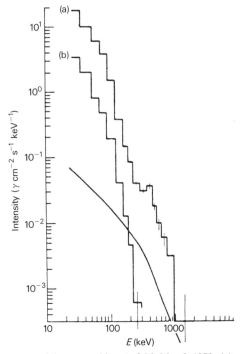

FIG. 10.30. Energy spectra of the unusual burst of 5th March 1979: (a) spectrum of the initial spike showing an emission line at $\varepsilon = 420$ keV; (b) spectrum of the oscillating component which followed the initial spike. The full line represents the spectrum of a normal burst. (From Mazets *et al.* 1979, p. 588.)

space density of $\sim 2.5 \times 10^{-2}$ pc^{-3} (Ostriker, Rees, and Silk 1970) the mean interval between bursts from a star would have to be less than $\sim 3 \times 10^5$ year; the mean interval could, of course, be much less than this if neutron stars are active over only a small fraction of their lifetime.

Several characteristics of the bursts imply that the dimensions of a source must be small. Intensity variations are seen on time scales of less than 0.1 s and, from considerations of the time for signals to cross the source, this sets an upper limit on the dimensions of $\sim 10^9$ cm. If it is assumed that the radiation is thermal, the measured spectra indicate that a typical temperature is $\sim 3 \times 10^8$ K; unless the optical depth of the source is very small, the luminosities which we have calculated imply that the dimensions are $\sim 10^5$ cm. Cavallo and Rees (1978) have discussed some of the general properties of sources, which they call cosmic fireballs, in which a very high energy density is suddenly created, initially in the form of gamma rays with energies of $\sim$ MeV. They showed that photon–photon collisions would create large numbers of electron–positron pairs and that Compton scattering would then establish an equipartition of energy between the photons and the particles. New photons are produced through bremsstrahlung from inter-actions between the fast electrons and the ambient gas; these photons then share the energy of the particles already present and the fireball cools. The spectrum of the radiation emerging from the fireball contains a high energy tail due to photons which escape from the surface of the fireball before interacting and redistributing their energy. The proportion of the radiation in this high energy tail decreases as the energy density in the source increases and Schmidt (1978) has shown that the intensities of $\sim$ MeV photons observed in bursts set an upper limit to the distances to the sources of a few kpc. Ramaty, Lingenfelter, and Bussard (1981) have extended this model to include the effects of synchrotron radiation, which would be important if the fireball were formed in the intense magnetic field at the surface of a neutron star. In a sufficiently strong magnetic field the electrons and positrons would lose energy more quickly by synchrotron radiation than by bremsstrahlung and this would lead to a faster cooling of the fireball. If the bursts come from fireballs formed on the surfaces of neutron stars the emission feature at $\varepsilon = 420$ keV could be annihilation radiation which had suffered a gravitational red shift when leaving the surfaces of the star. It remains a problem to explain how the initial energy of the fireball is created; suggested mechanisms include starquakes within the neutron star, the sudden accretion of a large mass of material on to its surface, or a thermonuclear explosion following a more gradual accretion.

References

Albats, P., Ball, S. E., Delvaille, J. P., Greisen, K. I., Koch, D. G., McBreen, B., Fazio, G. G., Hearn, D. R., and Helmken, H. F. (1971). *Nucl. Instrum. Methods* **95**, 189.

Baird, G. A., Delaney, T. J., Lawless, B. G., Griffiths, D. J., Shakeshaft, J. R., Drever, R. W. P., Meikle, W. P. S., Jelley, J. V., Charman, W. N., and Spencer, R. E. (1975). *Astrophys. J.* **196**, L11.

Bennett, K., Bignami, G. F., Boella, G., Buccheri, R., Hermsen, W., Kanbach, G., Lichti, G. G., Masnou, J. L., Mayer-Hasselwander, H. A., Paul, J. A., Scarsi, L., Swanenburg, B. N., Taylor, B. G., and Wills, R. D. (1977*b*). *Astron. Astrophys.* **61**, 279.

——, ——, Bonnardeau, M., Buccheri, R., Hermsen, W., Kanbach, G., Lichti, G. G., Mayer-Hasselwander, H. A., Paul, J. A., Scarsi, L., Stiglitz, R., Swanenburg, B. N., and Wills, R. D. (1977*a*). *Astron. Astrophys.* **56**, 469.

Bignami, G. F. and Morfill, G. E. (1980). *Astron. Astrophys.* **87**, 85.

Caraveo, P. A., Bennett, K., Bignami, G. F., Hermsen, W., Kanbach, G., Lebrun, F., Masnou, J. L., Mayer-Hasselwander, H. A., Paul, J. A., Sacco, B., Scarsi, L., Strong, A. W., Swanenburg, B. N., and Wills, R. D. (1980). *Astron. Astrophys.* **91**, L3.

Cavallo, G. and Jelley, J. V. (1975). *Astrophys. J.* **201**, L113.

—— and Rees, M. J. (1978). *Mon. Not. R. astr. Soc.* **183**, 359.

Cherry, M. L., Chupp, E. L., Dunphy, P. P., Forrest, D. J., and Ryan, J. M. (1980). *Astrophys. J.* **242**, 1257.

Chupp, E. L. (1976). *Gamma ray astronomy*. Reidel, Dordrecht.

Clayton, D. D. and Craddock, W. L. (1965). *Astrophys. J.* **142**, 189.

Cline, T. L. (1981). Proc. La Jolla Institute Workshop on Gamma-Ray Transients, 1981 August 5–8. (Preprint.)

—— and Desai, U. D. (1975). *Astrophys. J.* **196**, L43.

——, ——, Pizzichini, G., Teegarden, B. J., Evans, W. D., Klebsedal, R. W., Laros, J. G., Hurley, K., Niel, M., Vedrenne, G., Estulin, I. V., Kuznetsov, A. V., Zenchenko, V. M., Hovestadt, D., and Gloeckler, G. (1980). *Astrophys. J.* **237**, L1.

Evans, W. D., Klebsedal, R. W., Laros, J. G., Cline, T. L., Desai, U. D., Teegarden, B., Pizzichini, G., Margon, B., Hurley, K., Niel, M., Vedrenne, G., Estulin, I. V., Kuznetsov, A. V., and Zenchenko, V. M. (1980). *Astrophys. J.* **237**, L7.

Gold, T. (1969). *Nature (London)* **221**, 25.

Gould, R. J. (1965). *Phys. Rev. Lett.* **15**, 577.

Greisen, K. I. (1971). *The physics of cosmic X-ray, gamma ray and particle sources*, p. 84. Gordon and Breach, New York.

Grindlay, J. E., Helmken, H. F., and Weekes, T. C. (1976). *Astrophys. J.* **209**, 592.

Haymes, R. C., Walraven, G. D., Meegan, C. A., Hall, R. D., Djuth, F. T., and Shelton, D. H. (1975). *Astrophys. J.* **201**, 593.

Hermsen, W., Swanenburg, B. N., Bignami, G. F., Boella, G., Buccheri, R., Scarsi, L., Kanbach, G., Mayer-Hasselwander, H. A., Masnou, J. L., Paul, J. A., Bennett, K., Higdon, J. C., Lichti, G. G., Taylor, B. G., and Wills, R. D. (1977). *Nature (London)* **269**, 494.

Higdon, J. C. and Lingenfelter, R. E. (1976). *Astrophys. J.* **208**, L107.

Hillier, R. R., Jackson, W. R., Murray, A., Redfern, R. M., and Sale, R. G. (1970). *Astrophys. J.* **162**, L177.

Jennings, M. C. and White, R. S. (1980). *Astrophys. J.* **238**, 110.

Johnson, W. N. and Haymes, R. C. (1973). *Astrophys. J.* **184**, 103.

Kanbach, G., Bennett, K., Bignami, G. F., Buccheri, R., Carveo, P., D'Amico, N.,

Hermsen, W., Lichti, G. G., Masnou, J. L., Mayer-Hasselwander, H. A., Paul, J. A., Sacco, B., Swanenburg, B. N., and Wills, R. D. (1980). *Astron. Astrophys.* **90**, 165.

Klebsedal, R. W., Strong, I. B., and Olson, R. A. (1973). *Astrophys. J.* **182**, L85.

Leventhal, M. (1973). *Astrophys. J.* **183**, L147.

——, MacCallum, C. J., and Watts, A. C. (1977). *Astrophys. J.* **216**, 491.

——, ——, Huters, A. F., and Strang, P. D. (1980). *Astrophys. J.* **240**, 338.

——, ——, and Stang, P. D. (1978). *Astrophys. J.* **225**, L11.

McBreen, B., Ball, S. E., Campbell, M., Greisen, K., and Koch, D. (1973). *Astrophys. J.* **184**, 571.

Mayer-Hasselwander, H. A., Bennett, K., Bignami, G. F., Buccheri, R., D'Amico, N., Hermsen, W., Kanbach, G., Lebrun, F., Lichti, G. G., Masnou, J. L., Paul, J. A., Pinkau, K., Scarsi, L., Swanenburg, B. N., and Wills, R. D. (1980). *Proc. 9th Texas Symp. on Relativistic Astrophysics.* In *Ann. N.Y. Acad. Sci.* **336**, 211.

Mazets, E. P., and Golenetskii, S. U. (1981). *Astrophys. Space Sci.* **75**, 47.

——, ——, Aptekar, R. L., Guryan, Y. A., and Ilyinskii, V. N. (1981). *Nature (London)* **290**, 378.

——, ——, Ilyinskii, V. N., Aptekar, R. L., and Guryan, Y. A. (1979). *Nature (London)* **282**, 587.

——, ——, ——, Guryan, Y. A., Aptekar, R. L., Panov, V. N., Sokolov, I. A., Sokolova, Z. Y., and Kharitonova, T. V. (1982). *Astrophys. Space Sci.* **82**, 261.

Metzger, A. E., Parker, R. H., Gilman, D., Peterson, L. E., and Trombka, J. L. (1974). *Astrophys. J.* **194**, L19.

Ostriker, J. P., Rees, M. J., and Silk, J. (1970). *Astrophys. J.* **6**, 179.

Paciesas, W. S., Cline, T. L., Teegarden, B. J., Tueller, J., Durouchoux, P., and Hameury, J. M. (1982). *Astrophys. J.* **260**, L7.

Radhakrishnan, V. (1982). *Contemp. Phys.* **23**, 207.

Ramaty, R. and Linfenfelter, R. E. (1979). *Nature (London)* **278**, 127.

——, ——, and Bussard, R. W. (1981). *Astrophys. Space Sci.* **75**, 193.

Salvati, M. and Massaro, E. (1978). *Astron. Astrophys.* **67**, 55.

Schmidt, W. K. H. (1978). *Nature (London)* **271**, 526.

Strong, I. B., Klebsedal, R. W., and Olson, R. A. (1974). *Astrophys. J.* **188**, L1.

Sturrock, P. A. (1971). *Astrophys. J.* **164**, 529.

Swanenburg, B. N., Bennett, H. K., Bignami, G. F., Buccheri, R., Caraveo, P., Hermsen, W., Kanbach, G., Lichti, G. G., Masnou, J. L., Mayer-Hasselwander, H. A., Paul, J. A., Sacco, B., Scarsi, L., and Wills, R. D. (1981). *Astrophys. J.* **243**, L69.

Walraven, G. D., Hall, R. D., Meegan, C. A., Coleman, P. L., Shelton, D. H., and Haymes, R. C. (1975). *Astrophys. J.* **202**, 502.

Weisskopf, M. C., Silver, E. H., Kestenbaum, H. L., Long, K. S., and Novick, R. (1978). *Astrophys. J.* **220**, L117.

Wills, R. D., Bennett, K., Bignami, G. F., Buccheri, R., Carevo, P. A., Hermsen, W., Kanbach, G., Masnou, J. L., Mayer-Hasselwander, H. A., Paul, J. A., and Sacco, B. (1982). *Nature (London)* **296**, 723.

Wilson, A. S. (1972). *Mon. Not. R. astr. Soc.* **160**, 355.

11

EXTRAGALACTIC GAMMA RAY SOURCES AND THE DIFFUSE BACKGROUND RADIATION

11.1. Introduction

The detection by radioastronomers in the 1950s of strong synchrotron radiation from extragalactic sources showed that a small proportion of galaxies contain extremely high concentrations of relativistic particles, similar to those which constitute the cosmic rays in our own Galaxy but with a much greater intensity. Since many of the mechanisms for the generation of gamma rays involve the interactions of relativistic particles it is natural to expect these galaxies to be strong sources of gamma rays. Moreover the radio observations indicate that the luminosities of these extragalactic sources may be so high that, despite their very great distances from us, they may appear as some of the brightest sources in the sky. The bulk of the radioemission from one of these galaxies usually comes from two huge lobes on either side of the galaxy, but detailed mapping often reveals a very compact source at the nucleus of the galaxy; this compact source is probably the origin of the relativistic particles. There have been many speculations that the same type of compact source may account for the bright nuclei of Seyfert galaxies and, on a more powerful scale, for quasars. The physical conditions inside the compact sources, with their strong magnetic fields, high radiation densities, and large fluxes of relativistic particles, are particularly favourable for the generation of gamma rays.

11.2. Radiation processes in compact sources

Since the compact sources seem to be the origin of the relativistic particles which are seen elsewhere in the galaxies it is tempting to try to explain all the emissions observed from the compact sources in terms of interactions which the relativistic particles suffer before they escape. On this approach the radio and infrared emission would be interpreted as synchrotron radiation and the X-rays and gamma rays could be the result of Compton scattering of the synchrotron photons by the same electrons. This two-stage mechanism is usually referred to as the Compton–synchrotron process.

When constructing a model for a source of synchrotron radiation a common problem is the lack of a direct measurement of the strength of the magnetic field in the source. One way of overcoming this is to identify the frequency at which self-absorption of the synchrotron radiation occurs. A synchrotron source becomes optically thick (Tucker 1975) when its brightness temperature, $T_B(v)$, is given by

$$kT_B(v) \sim E,$$

where E is the energy of the relativistic electrons which generate synchrotron radiation at frequency v. The brightness temperature of the source is related to the luminosity, $L(v)$, and the radius, R, by

$$L(v) = 2\pi k T_B(v) \frac{v^2}{c^2} \pi R^2,$$

i.e. $T_B(v)$ is the temperature of the black body which would have the same luminosity at this frequency. The flux, $j(v)$, received at the earth is

$$j(v) = \frac{L(v)}{4\pi r^2} = \frac{kT_B}{2\pi} \frac{v^2}{c^2} \frac{R^2}{r^2}$$

where r is the distance to the source. Now, $R/r = \theta$, the angular size of the source, and the flux at the frequency, v_a, where self-absorption sets in is given by

$$j(v_a) \propto E v_a^2 \theta^2.$$

In Section 2.4 we saw that

$$v \propto E^2 H,$$

so

$$j(v_a) \propto v_a^{5/2} H^{-1/2} \theta^2. \tag{11.1}$$

If θ is known, then H can be calculated.

In cases where the angular diameter of the source is not known, but where the emission from the Compton–synchrotron process can be measured, another approach is possible. Consider the synchrotron radiation from an assembly of electrons with an energy spectrum of the form

$$n(E)\, dE \propto E^{-m}\, dE.$$

The spectrum of the synchrotron radiation from these electrons is given by (Section 2.4),

$$j_s(v)\, dv \propto v^{-\alpha}\, dv$$

where $\alpha = (m - 1)/2$.

Jones, O'Dell, and Stein (1974) have shown that the flux, j_{cs}, of Compton–

synchrotron radiation from the electrons is given by

$$j_{cs}(v) \propto j_s(v) v^{-m} H^{m-1} \theta^{-2}. \tag{11.2}$$

Equations 11.1 and 11.2 provide two relationships from which θ and H can be calculated.

11.3. Seyfert galaxies

11.3.1. *Introduction*

A review of the properties of Seyfert galaxies has been given by Weedman (1977). Visually, a Seyfert galaxy is easily identified by its bright starlike nucleus which is unresolved in ground-based optical telescopes. Photographs of the nearby galaxy NGC 4151 taken from a balloon show that its nucleus has an angular diameter of less than 0.1 arc s which corresponds to a linear diameter of less than 10 pc. The nuclei of several Seyfert galaxies show large irregular variations in luminosity over time scales of the order of months which implies dimensions of less than $\sim 10^{-1}$ pc.

The optical spectrum of the nucleus of a Seyfert galaxy is unlike that of a normal galaxy, with strong broad emission lines. On the basis of their emission line spectra Seyfert galaxies have been classified into two types. In Type 1 galaxies the Balmer lines of hydrogen, with a width corresponding to a velocity dispersion of $\sim 10^4$ km s^{-1}, are an order of magnitude broader than the other bright emission lines, the forbidden lines of oxygen and nitrogen; by contrast, in Type 2 galaxies both the Balmer lines and the forbidden lines have a width of $\sim 10^3$ km s^{-1}. The continuous spectra of the two types are also different with Type 2 galaxies having thermal spectra and Type 1 galaxies having spectra in the form of a power law which gives rise to anomalously high emissions in the ultraviolet and the infrared.

Many Seyfert Type 1 galaxies are known to be strong sources of X-rays. The X-ray luminosity tends to be correlated with the optical luminosity of the galactic nucleus and this correlation, together with the large time variations in the X-ray emission, suggests that the X-rays originate in the nuclei. The X-ray spectrum can usually be represented by

$$j(v) \, dv \propto v^{-\alpha} \, dv$$

with α in the range ~ 0.3 to ~ 1.0. With one or two exceptions the spectra show no evidence for absorption at low energies and this places upper limits of $\sim 5 \times 10^{22}$ atoms cm^{-2} on the surface densities of the sources.

11.3.2. *Gamma ray emission from NGC 4151*

NGC 4151, at a distance of only 19 Mpc, is one of the closest and the most studied of the Seyfert Type 1 galaxies. Like most Seyfert galaxies the strength of its radio emission resembles that of a normal galaxy, but the infrared emission from its nucleus is unusually large and variable.

The X-ray spectrum has the form of a power law

$$n(\varepsilon)\, d\varepsilon = \frac{j(\varepsilon)\, d\varepsilon}{\varepsilon} \propto \varepsilon^{-(\alpha+1)}\, d\varepsilon$$

up to photon energies of $\sim 200\,\mathrm{keV}$ (Auriemma *et al.* 1978). The X-ray emission is highly variable and a flare lasting only ~ 1.5 days has been observed (Mushotzky, Holt, and Serlemitsos 1978*a*). The measured values of the spectral index, α, range from ~ 0.0 to ~ 0.6, with a mean value of ~ 0.3. NGC 4151 is unusual amongst the Seyfert galaxies showing X-ray emission in that its spectrum has evidence of absorption at photon energies below a

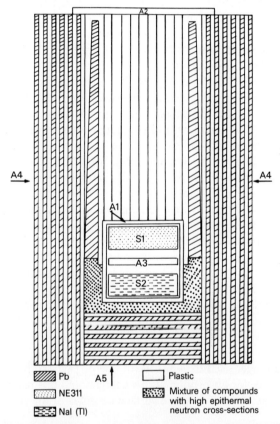

FIG. 11.1. Diagram of the MISO low energy gamma ray telescope. The central detectors S1 and S2 could be operated either singly as two separate detectors or in coincidence as a Compton telescope. The detectors A1, A2, A3, and A4 were anticoincidence counters. (From Baker *et al.* 1979, p. 596.)

few eV, which implies a surface density for the source of $\sim 10^{23}$ atoms cm^{-2}. If this absorption occurs in an interstellar medium similar to that in our own Galaxy then scattering by dust should lead to considerable reddening of the optical radiation, but this is not observed. This conflict may be resolved by assuming that the optical radiation originates in a larger region outside the X-ray source and does not therefore have to pass through the same column density of material.

Measurements of the low energy gamma ray flux from NGC 4151 indicate that the power law spectrum may extend to photon energies of ~ 5 MeV. Perotti et al. (1979) have reported measurements made from a balloon in May 1977. The telescope (Fig. 11.1) had a sensitive area of 500 cm^2 and consisted of a liquid scintillation counter and a sodium iodide scintillation counter; the counters could be operated either singly as two separate detectors or in coincidence as a Compton telescope. Both detectors were inside a massive shield consisting of interleaved layers of plastic scintillator and lead, and the lower detector was also surrounded by wax loaded with compounds of boron, lithium, and tungsten—elements which have a high cross-section for thermal and epithermal neutrons. Measurements on NGC 4151 were made at an atmospheric depth of ~ 7 g cm^{-2} for a period of 3 h. An excess counting rate, amounting to ~ 3 per cent of the background, was observed from the direction of NGC 4151, the excess being most significant when the detectors were operated singly rather than as a Compton telescope. The flux measured by this experiment can be reconciled with upper limits to the flux of low energy gamma rays set in October 1977 by Meegan and Haymes (1979) and in September 1978 by White et al. (1980) only by assuming that the source is highly variable on time scales of the order of a year. At higher photon energies the experiments on the SAS-2 and COS-B satellites have only been able to set upper limits to the flux. The X-ray and gamma ray spectrum from NGC 4151 is shown in Fig. 11.2; it is clear that the power law spectrum

$$n(\varepsilon)\, d\varepsilon \propto \varepsilon^{-(\alpha+1)}\, d\varepsilon$$

with $\alpha \sim 0.3$ does not extend to photon energies greater than a few MeV.

Mushotzky (1977) has proposed a model for the nucleus in which the radiation is produced by the Compton–synchrotron process described in Section 11.2. According to this model the emission at millimetre wavelengths is synchrotron radiation from relativistic electrons moving in a magnetic field of a few gauss; some of these photons undergo Compton scattering with the same electrons, producing the X-ray emission. To account for the variability of the source it is assumed that the electrons are injected in short bursts, with long intervals in between. In Chapter 2, it was shown that, if the spectrum of the electrons has the form

$$n(E)\, dE \sim E^{-m}\, dE,$$

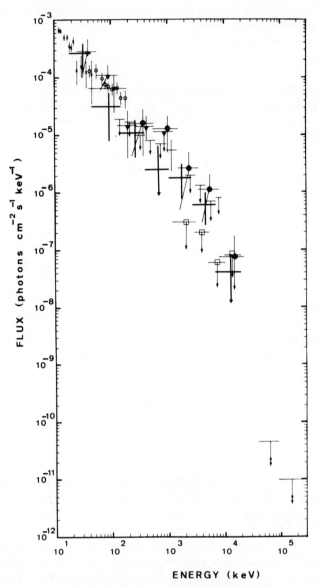

FIG. 11.2. X-ray and gamma ray spectrum of NGC 4151. (From Perotti *et al.* 1981.)

then the synchrotron radiation will have a spectrum given by

$$j(v)\, dv \sim v^{-\alpha}\, dv$$

where $\alpha = \frac{1}{2}(m - 1)$.

The synchrotron (and the Compton) energy losses are most severe for the higher energy electrons and this causes the energy spectrum of the electrons to steepen with time. The synchrotron spectrum also steepens and the spectral index becomes (Tucker 1975),

$$\alpha' = \frac{1}{3}(4\alpha + 3).$$

This break in the synchrotron spectrum should be reproduced, through the subsequent Compton scattering of some of the photons, in the X-ray spectrum from the source.

Measurements of NCG 4151, made over the full range of the electromagnetic spectrum, are shown in Fig. 11.3 which indicates that the luminosity of the source is dominated by the features at infrared and X-ray frequencies. The infrared emission is probably complex with a large thermal contribution from stars and dust, but there is also evidence (Cutri and Rudi 1980) for a non-thermal component from the nucleus of the galaxy with a spectral index $\alpha = 0.4$. In Mushotzky's model the break in the synchrotron spectrum occurs between infra-red and optical frequencies and the corresponding break in the Compton spectrum occurs at photon energies of

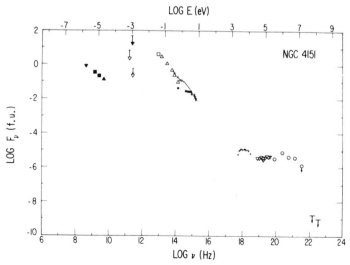

FIG. 11.3. Measurements of the radiation from NGC 4151 over the entire electromagnetic spectrum. (From Schlickeiser 1980, p. 640.) © (1980) The American Astronomical Society.

~ 1 MeV. This explains why the spectral indices are similar at infrared and X-ray frequencies, and suggests that the high energy gamma ray spectrum should have an index

$$\alpha' = \tfrac{1}{3}(4\alpha + 3) \sim 1.5,$$

which is consistent with the data from the SAS-2 and COS-B experiments. Support for the Compton–synchrotron model of the emission from the nuclei of Seyfert galaxies comes from the correlation observed between the infrared and the X-ray intensities measured from a sample of 10 galaxies (Elvis *et al.* 1978).

An alternative model for the source in NGC 4151, in which the emission comes from a hot electron gas in an accretion disc around a massive body at the centre of the galaxy, has been suggested by Katz (1976). The bremsstrahlung radiation from a hot electron gas has a Wien-spectrum of the form

$$j(v)\,\mathrm{d}v \propto v^2 \exp\left(-\frac{hv}{kT}\right)\mathrm{d}v,$$

which does not resemble the spectrum observed from NGC 4151. Katz therefore assumed that the electrons produce the radiation by multiple Compton scattering of photons which originate in an unidentified source. This mechanism, which has already been invoked to explain the radiation from stellar X-ray sources such as Cygnus X-1, produces a spectrum which approximates to a power law up to a photon energy of $\sim 2kT$, where T is the temperature of the electron gas. There is ample evidence that accretion discs exist in many close binary systems of stars but there is little direct evidence that similar discs exist on a galactic scale. It has been suggested that the broad hydrogen lines which are characteristic of a Type 1 Seyfert galaxy may originate in a gaseous disc, the width of the lines being caused by the rotation of the disc. One general argument pointing to the existence of accretion discs at the nuclei of galaxies is that the axial symmetry of a disc would provide a basis for an explanation of the linear structures, such as jets and lobes, which are frequently associated with active galaxies.

11.4. Radio galaxies

11.4.1. *Introduction*

Radio galaxies were originally defined as those galaxies, usually elliptical in shape, whose radio emission was several orders of magnitude greater than that of a normal galaxy. As the angular resolution of radiotelescopes improved it became apparent that the radio emission usually comes from two lobes on either side of the galaxy. More recently, using arrays of telescopes with angular resolutions better than a milliarcsecond, it has been shown that most, if not all, of these galaxies have a compact source at the nucleus often with a linear feature, or jet, emerging from it and aligned with the lobes.

The spectrum and polarization of the radiation from the lobes identifies it as synchrotron radiation. The origin and the acceleration mechanism of the relativistic electrons is still unknown but the morphology of the galaxy points to the compact nuclear object as the source of the energy.

11.4.2. *X-ray and gamma ray emission from the radio galaxy Centaurus A (NGC 5128)*

NGC 5128, at a distance of 5 Mpc, is one of the nearest radio galaxies and its lobes are separated in the sky by $\sim 5°$. There is a compact source, with an angular diameter of less than a milliarcsecond, in its nucleus and this source is highly luminous in the millimetre radio and infrared regions of the spectrum. The X-ray emission from NGC 5128 has been shown (Schreier *et al.* 1979) to originate principally from an unresolved source in the nucleus of the galaxy. The X-ray spectrum can be fitted to a power law with an index $\alpha \sim 0.7$; there is evidence of absorption at low energies implying a column density in the source of 1.3×10^{23} atoms cm^{-2}. Large variations in the X-ray emission have been observed (Beall *et al.* 1978) with increases by a factor of ~ 4 occurring over a period of a year.

Evidence that the power law spectrum extends to the low energy gamma ray region of the spectrum was provided by a balloon experiment (Hall *et al.* 1976) carried out by a team from Rice University using the detector described in Section 10.2. The measurements were made in April 1974 for a period of 3 h at an atmospheric depth of $4 \, g \, cm^{-2}$; the spectrum measured from NGC 5128 is shown in Fig. 11.4. The authors claimed evidence for nuclear lines in the spectrum but the data were of low statistical weight and have not been confirmed. Indeed, no gamma ray flux, neither lines nor continuum, had been detected by the Rice University group when they flew a similar instrument on a balloon in May 1968; it may be significant that the X-ray flux was observed to increase by a factor of ~ 4 between May 1968 and April 1974. Measurements with the SAS-2 and COS-B satellites yielded only upper limits to the flux of high energy gamma rays. Thus the gamma ray observations of NGC 5128, like those of NGC 4151, are confined to energies below a few MeV, are of low statistical weight, and require a variable flux if the results of all the measurements are to be reconciled. However, the upper limits to the flux of high energy gamma rays indicate that the power law spectrum observed in the X-ray region does not extend to gamma ray energies greater than 30 MeV. The consequent curvature in the spectra will be of importance when we consider, in a later section, the contribution which these sources make to the cosmic background radiation.

NGC 5128 is one of the few sources from which the detection of very high energy gamma rays has been claimed. Grindlay, Helmken, Hanbury-Brown,

Davis, and Allen (1975) reported measurements made using a ground-based detecting system at Narrabi, N.S.W., to observe the Cerenkov light from electromagnetic showers produced by gamma rays in the atmosphere as described in Chapter 7. The detectors were photomultipliers placed at the face of two large optical reflectors. The reflectors were separated by a distance of 120 m and parallax made it possible to distinguish electro-magnetic showers produced by gamma rays from those produced by cosmic ray nuclei because the latter reach a maximum intensity at a lower depth in the atmosphere. Measurements were made on many nights in 1972, 1973, and 1974. A small signal, which was only ~ 1 per cent of the background flux, was detected each year; the signal was smallest in 1972 when the microwave emission from NGC 5128 was also low. The combined results give a signal which was 4.5 standard deviations above the background intensity. This would usually be regarded as a significant result but experience has shown that large unexplained fluctuations do sometimes occur in air shower experiments. This signal would correspond to an integral flux of 4×10^{-11} photons cm^{-2} s^{-1} at energies greater than 3×10^5 MeV.

Measurements of the spectrum of the nucleus of NGC 5128 over the full range of frequency is shown in Fig. 11.5. Several models have been proposed to explain this spectrum; most models consider the radio emission to be synchrotron radiation and the X-rays and gamma rays to be Compton

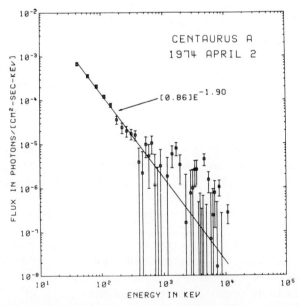

FIG. 11.4. Spectrum of NGC 5128 measured with the balloon-borne telescope built at Rice University. (From Hall *et al.* 1976, p. 637.) © (1976) The American Astronomical Society.

radiation from the same relativistic electrons because this accounts naturally for the similarity of the spectral indices in these two regions of the spectrum. The models differ in the origin they assume for the low energy photons required in the Compton process. Grindlay (1975) and Mushotzky, Serlemitsos, Becker, Boldt, and Holt (1978b) take the low energy photons to be those generated in the synchrotron process whereas Beall et al. (1978) postulate an independent source of black-body radiation within the nucleus which is not directly observable owing to obscuration by dust.

If the gamma rays are produced by the Compton–synchrotron process, then the spectrum shown in Fig. 11.5 allows us to estimate the two parameters which we saw, in Section 11.2, were required in an analysis of such a source; these parameters are the frequency at which self-absorption of the synchrotron radiation takes place and the ratio of the intensity of synchrotron radiation to that of Compton radiation from the same electrons. Using the arguments developed in Section 11.2 we calculate the diameter of the source in NGC 5128 to be ~7 light days and the magnetic field strength to be ~0.1 G. The variable luminosity of the source implies that the relativistic electrons are injected in bursts. A problem with this model is that the X-ray luminosity has been observed to vary on a time scale much shorter than that over which the electrons lose their energy by radiation, which can be calculated from the parameters derived above to be several months.

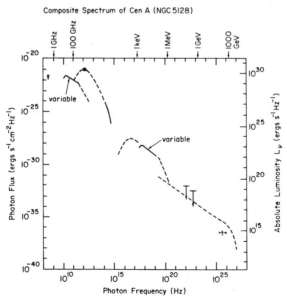

FIG. 11.5. Measurements of the radiation from NGC 5128 over the entire electromagnetic spectrum. (From Beall 1979, p. 149.)

Mushotzky *et al.* (1978*b*) therefore proposed that the energy losses of the electrons are dominated by adiabatic expansion of the source rather than by radiation.

11.5. Quasars

11.5.1. *Introduction*

Quasars are observed as strong radio sources, often with the double-lobe structure characteristic of radio galaxies. Optically, however, they appear as starlike images and their optical spectra contain the broad emission lines associated with the nuclei of Seyfert galaxies. The spectral lines have large red shifts and there has been a long debate as to the origin of this effect. There is now a wide consensus that the red shifts are of cosmological origin and this places the quasars at very great distances and implies that, despite their small size, they are amongst the most luminous objects in the universe. Quasars resemble, in many ways, the nuclei of Type 1 Seyfert galaxies; their line spectra contain broad emission lines of hydrogen, their optical continuous spectra usually take the form of a power law, and their luminosities show large variations on time scales as short as days. However, the luminosities of quasars are up to four orders of magnitude greater than those of Seyfert galaxies and their large radio lobes indicate that, unlike Seyfert galaxies, they eject large quantities of energy in the forms of relativistic particles and magnetic fields.

11.5.2. *The X-ray and gamma ray emission from the quasar 3C273*

Optically 3C273 has the greatest apparent brightness of all the quasars and its red shift of $z = \Delta\lambda/\lambda = 0.158$ places it at a distance of 800 Mpc. Using instruments on the HEAO-1 satellite Primini *et al.* (1979) detected X-rays from 3C273 at photon energies up to 120 keV. The spectrum could be represented by a power law

$$n(\varepsilon)\,\mathrm{d}\varepsilon \propto \varepsilon^{-(\alpha+1)}\,\mathrm{d}\varepsilon$$

with $\alpha = 0.67 \pm 0.14$. Measurements with other detectors on the same satellite showed that the soft X-ray intensity increased by ~ 40 per cent between December 1977 and June 1978, and a 10 per cent variation in the X-ray intensity was detected over a period of 15 h implying an upper limit to the dimensions of the source of $\sim 1.5 \times 10^{15}$ cm. There is no evidence of absorption in the soft X-ray spectrum which indicates that the column density in the source is less than 5×10^{21} atoms cm^{-2}, much smaller than that in the source in NGC 5128.

It has only been possible to set upper limits to the low energy gamma ray flux from 3C273, but Bignami *et al.* (1981) measured the high energy gamma ray flux using the spark chamber on the COS-B satellite. The spectrum, if expressed in the form of a power law, had a spectral index $\alpha = 1.6 \pm 0.4$. The X-ray and gamma ray measurements are shown in Fig. 11.6 and the break in the spectrum is clearly evident.

There is a fundamental problem associated with any model for 3C273 because of its high luminosity and its small dimensions. McBreen (1979) has pointed out that an X-ray source can become opaque to high energy gamma rays through the process

$$\gamma_1 + \gamma_2 \rightarrow e^+ + e^-.$$

The threshold energy for this interaction is given by

$$\varepsilon_1 = \frac{2m^2c^4}{\varepsilon_2(1 - \cos\theta)}$$

where ε_1 and ε_2 are the energies of the two photons and θ is the angle between

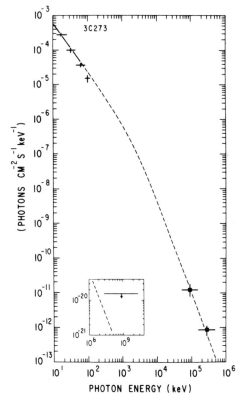

FIG. 11.6. X-ray and gamma ray spectrum of 3C273. The insert shows an upper limit to the flux of very high energy gamma rays set by ground-based measurements. (From Bignami, Fichtel, Hartman, and Thompson 1979, p. 652.)

their directions of motion. For an isotropic flux of radiation we can write

$$\langle 1 - \cos\theta \rangle = 1.$$

The optical depth for gamma rays is given by

$$\tau \sim r \int_{\varepsilon_1}^{\infty} n(\varepsilon)\sigma(\varepsilon)$$

where $\sigma(\varepsilon)$ is the interaction cross-section and $n(\varepsilon)$ is the number density of target photons in the source. The cross-section rises rapidly with energy, reaches a maximum of 1.7×10^{25} cm^2 at

$$\varepsilon_{max} \sim 2\varepsilon_{threshold},$$

and then falls as ε^{-2} for energies well above threshold.

Using an upper limit to the size of the source in 3C273 of 1.5×10^{15} cm, as implied by the X-ray variability, we find that the source is optically thick for gamma rays with energies above a few MeV. This result leads to the conclusion that the X-rays and high energy gamma rays must come from different regions of the source; one way of avoiding this restriction is to assume that both forms of radiation are strongly beamed, so that θ is small, and the threshold energy for the interaction is correspondingly high.

11.6. Cosmological background radiation

11.6.1. Introduction

The problem of distinguishing primary gamma rays from locally produced background radiation, which was discussed in Chapter 3, is nowhere more apparent than in the experiments to detect an isotropic component of cosmic gamma rays. Unlike the fluxes from discrete sources, a diffuse flux cannot easily be distinguished from local background radiation merely by its angular distribution since many detectors do not distinguish upward-moving from downward-moving gamma rays. Moreover experiments in which attempts are made to identify a primary flux by modulating it with a system of moving absorbers are difficult to interpret because the absorbers can also act as additional sources of background radiation. As a result of these problems, despite many attempts to measure a diffuse primary flux and although the flux may well be stronger than that from discrete sources, there is still great uncertainty as to its intensity and spectrum. For this reason it is necessary to look even more closely than usual at the integrity of the measurements.

11.6.2. The diffuse X-ray flux

We saw in Chapter 3 that the spectrum of the background radiation produced in the earth's atmosphere has a maximum at X-ray energies of

~ 50 keV. At energies well below this maximum the primary flux at the top of the atmosphere is considerably stronger than the locally produced background. The existence of the primary flux is then evident in the data from a simple detector and collimator which rotates and scans the sky and the atmosphere, because the intensity of the diffuse radiation is higher when the detector points upwards at the sky rather than when it points downwards at the atmosphere. Marshall *et al.* (1980) made measurements of the diffuse flux at X-ray energies from 3 keV to 50 keV using a detector designed specifically for this purpose on the HEAO 1 satellite. The spectrum could be fitted very well by the thermal bremsstrahlung from a hot gas with a temperature of 5×10^8 K.

At X-ray energies above ~ 30 keV the problem posed by the locally produced background radiation becomes apparent. One way of disentangling the two components is to record the total flux as a function of depth in the atmosphere and then to use a model for the production of the atmospheric radiation, which can be fitted at large depths, to calculate the local contribution to the total flux at the top of the atmosphere. Figure 11.7 shows the counting rates recorded in a scintillation counter, fitted with a vertical collimator, which were measured during the ascent of a balloon. Very little primary flux penetrates to depths greater than ~ 12 g cm^{-2} and the total counting rate below that depth shows the locally produced background radiation increasing with depth. At depths less than 12 g cm^{-2} the contribution from the primary flux causes the total counting rate to decrease with depth; this change in the slope of the transition curve at small depths is an unambiguous signature of a primary flux and data such as that shown in Fig. 11.7 can be used, with some confidence, to estimate the primary flux for X-ray energies up to ~ 70 keV. At higher energies, however, the results are too dependent on the model assumed for the production of the atmospheric radiation. One way of checking that the contribution from the atmospheric background has been correctly allowed for is to make measurements at different geomagnetic latitudes, since the primary flux should be independent of latitude although the atmospheric background will vary considerably.

11.6.3. *The diffuse flux of low energy gamma rays*

In this region of the spectrum the intensity of the locally produced radiation is very much greater than that of the primary radiation and the total counting rate in any detector, even if it records only downward-moving photons, decreases monotonically to the top of the atmosphere. A simple model for the production of the atmospheric radiation will illustrate the problem.

Let the production rate of background radiation per unit mass of the atmosphere be b photons g^{-1} s^{-1} sr^{-1}. We shall assume this to be independent of depth although in a more realistic model it should increase in proportion to the local intensity of the nuclear component of the cosmic

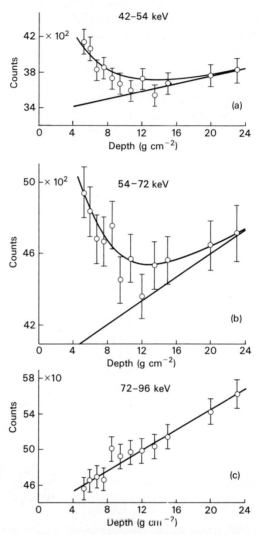

FIG. 11.7. Counting rates of a balloon-borne X-ray detector as a function of atmospheric depth: (a) $42 \text{ keV} \leqslant \varepsilon < 54 \text{ keV}$; (b) $54 \text{ keV} \leqslant \varepsilon < 72 \text{ keV}$; (c) $72 \text{ keV} \leqslant \varepsilon \leqslant 96 \text{ keV}$. Note the suppressed zeros on the abscissa scales. The straight lines represent estimates of the atmospheric background flux deduced from measurements at depths greater than $\sim 20 \text{ g cm}^{-2}$. The curved lines in (a) and (b) are the transition curves expected for an isotropic primary flux normalized to the measured intensity at $\sim 5 \text{ g cm}^{-2}$. The error bars are $\pm 2\sigma$. (From Bridgeland 1972.)

radiation. Consider the counting rate of a telescope, with an effective area A, which accepts downward-moving photons within a small solid angle Ω about the vertical direction. If the primary flux is S_0 photons cm^{-2} s^{-1} sr^{-1}, then the counting rate $c(x)$ at a depth x in the atmosphere will be given by

$$c(x) = A\Omega\left(S_0\, e^{-\mu x} + \int_x^0 b\, e^{-\mu x}\, \mathrm{d}x \right)$$

$$= A\Omega\left\{ S_0\, e^{-\mu x} + \frac{b}{\mu}(1 - e^{-\mu x}) \right\}$$

$$= A\Omega\left\{ \frac{b}{\mu} + \left(S_0 - \frac{b}{\mu} \right) e^{-\mu x} \right\}.$$

Thus, if $b/\mu > S_0$ the counting rate will decrease monotonically with altitude and any attempt to measure the primary flux from an analysis of the counting rate as a function of altitude will depend critically on the model adopted for the production of the atmospheric radiation. In particular the extrapolation to a finite counting rate at zero atmospheric depth, which can be seen in the transition curves for X-ray energies in Fig. 11.8(c), should not be taken as evidence for a primary flux since all detectors are subject to background counting rates which are not caused by photons moving within the acceptance angle of the collimator and which therefore do not extrapolate to zero at the top of the atmosphere.

One way of overcoming the large flux of background radiation from the atmosphere is to make measurements at a point in deep space where the flux from the earth has been attenuated by the inverse-square law. Trombka et al. (1977) used this approach on the Apollo 15 and Apollo 16 manned missions to the Moon, the measurements being made when the spacecraft was in cislunar space. The detector was a sodium iodide scintillation crystal, 7 cm in diameter × 7 cm thick; to reject events caused by charged particles the crystal was surrounded by a plastic scintillator 1 cm thick which was viewed by a separate photomultiplier. The detector was mounted on a boom which, when fully extended, placed the detector at a distance of 7.6 m from the spacecraft; in this position the spacecraft subtended a solid angle of only 0.28 sr at the detector.

The measurements were made over a period of 3 days when the solid angles subtended by the earth and the moon were both less than 0.02 sr. Pulse height spectra were recorded for several positions of the boom. When the detector was close to the spacecraft the spectra contained several nuclear lines but, with the exception of the annihilation line at 0.51 MeV, these all disappeared when the boom was extended. The overall counting rate, corresponding to energy losses in the crystal between 0.3 MeV and 10 MeV,

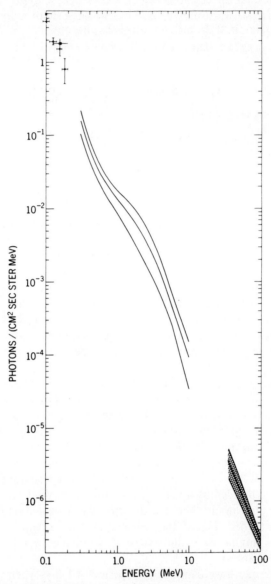

FIG. 11.8. Energy spectrum of the extragalactic diffuse gamma ray flux. The spectrum from 0.3 MeV to 10 MeV was calculated from data measured on the Apollo 15 and Apollo 16 spacecraft. The spectrum from 30 MeV to 100 MeV was measured by the spark chamber on the SAS-2 satellite. In each case the central line is the best estimate of the spectrum and the outer lines are one standard deviation limits. (From Trombka *et al.* 1977, p. 932.) © (1977) The American Astronomical Society.

decreased by a factor of only ~ 5 when the boom was fully extended although the solid angle subtended by the spacecraft decreased by a factor of ~ 20. The authors concluded that less than ~ 20 per cent of the counting rate when the boom was fully extended was due to background radiation produced in the spacecraft.

Dyer, Engel, and Quenby (1972) have pointed out that the technique of extending the detector on a long boom, although it allows a correction to be made for radiation from the spacecraft, does not overcome the problem of background effects produced in the detector itself. Of particular importance is the contribution from radioactivity which is induced when nuclei in the detector undergo spallation by cosmic ray particles; the decay of nuclear fragments may take place long after the passage of the cosmic ray particle and the signal from the radioactive decay cannot be vetoed by normal anticoincidence techniques. Dyer *et al.* showed that this effect may seriously contaminate the data at photon energies of a few MeV and it is suspicious that a feature appears in the measured spectrum at these energies.

Despite the problem of induced radioactivity the Apollo measurements shown in Fig. 11.8 represent probably the best estimate we have of the diffuse flux of low energy gamma rays; however, it may be prudent to regard the measurements as only upper limits to the true flux.

11.6.4. *The diffuse flux of high energy gamma rays*

The problem of distinguishing a primary flux from local background is somewhat easier in measurements with spark chambers at photon energies above ~ 30 MeV because the characteristic tracks of an electron–positron pair in a chamber ensure that the event is due to a gamma ray and also allow the direction of the gamma ray to be determined.

We saw in Chapter 9 that the Galactic disc is a strong source of high energy gamma rays and care must be taken to differentiate this Galactic radiation from a diffuse extragalactic flux. One way of doing this is to compare the gamma ray flux with the intensity of the radio emission at a wavelength of 21 cm at different Galactic latitudes. The radio emission defines the column density, N_H, of interstellar hydrogen atoms in a particular direction and since, as we have seen in Chapter 9, the high energy gamma rays are probably produced in cosmic ray interactions with the interstellar gas, we should be able to express the observed gamma ray flux, $j(\varepsilon)$, as

$$j(\varepsilon) = A(\varepsilon)N_H + B(\varepsilon)$$

where $B(\varepsilon)$ is the extragalactic component. Figure 11.9 shows the data from the spark chamber experiment on the SAS-2 satellite plotted in this way, and there is clear evidence for a component of the diffuse flux whose intensity is not correlated with the interstellar gas and which is therefore assumed to be extragalactic. The spectrum of this component, if it is expressed as a power law

$$n(\varepsilon)\, d\varepsilon \propto \varepsilon^{-(\alpha+1)}\, d\varepsilon,$$

has a spectral index of $\alpha = 1.7 \pm 0.4$. The Galactic component of the flux has a spectral index of $\alpha = 0.5 \pm 0.3$.

The statistical weight of the data is low but Fichtel, Simpson, and

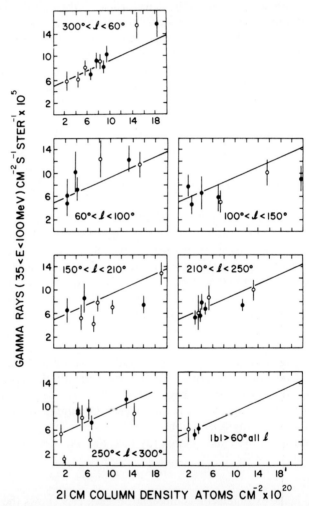

FIG. 11.9. Intensities of the diffuse flux of high energy gamma rays measured in different directions by the experiment on the SAS-2 satellite, plotted against the hydrogen column density as reduced from measurements of the 21 cm radio emission. The solid line is the best fit to all the data. (From Fichtel *et al.* 1978, p. 837.) © (1978) The American Astronomical Society.

Thompson (1978) have attempted to examine the isotropy of the extra-galactic component by comparing the intensities in two intervals of Galactic longitude, $300° < l < 60°$ and $100° < l < 250°$. The ratio of the intensities in the two intervals was 1.10 ± 0.19. Similarly the ratio of the intensities in the two intervals of Galactic latitude, $20° < b < 40°$ and $b > 60°$, was 0.87 ± 0.09. Thus the evidence is not inconsistent with the assumption that the extragalactic flux is isotropic, and is therefore probably cosmological in origin.

11.6.5. Origin of the diffuse extragalactic radiation

Theories of the origin of the diffuse extragalactic radiation fall into two classes: one class assumes that the radiation was produced in intergalactic space, usually at an earlier epoch in the universe, and is therefore genuinely isotropic whilst the other class assumes that the radiation is really the integrated flux from a large number of discrete sources and that the angular distribution of the radiation should therefore exhibit fluctuations when examined on a sufficiently fine scale. Both classes of theories run into difficulties because a large proportion of the flux inevitably originated at a much earlier epoch in the universe when conditions were very different from the present and about which we have very little other information.

There is some evidence that the X-rays and the gamma rays have quite distinct origins. The spectrum measured by the HEAO-1 experiment suggests that the X-rays are thermal bremsstrahlung from a plasma with a tempera-ture of 5×10^8 K. There are no known discrete extragalactic sources with similar thermal spectra, but to heat any intergalactic gas to this temperature would require a prohibitive energy source, so Boldt (1980) has proposed that the X-rays come from an as yet undiscovered class of sources which are all at very great distances and may have existed only at an earlier epoch in the universe.

When the thermal X-ray component is subtracted from the measured spectrum, the component which remains has a spectrum which is flat up to photon energies of a few MeV and then steepens. This resembles the spectra observed from active galaxies such as NGC 4151 and 3C273 and we shall consider whether sources such as these are capable of producing the gamma ray component of the diffuse flux.

Consider first a Euclidean universe with a uniform distribution of sources, each with a luminosity $L(\varepsilon)$. The total energy flux at any point is given by

$$j(\varepsilon) = \frac{1}{4\pi} \int_0^{r_{max}} 4\pi r^2 n \frac{L(\varepsilon)}{4\pi r^2} \, dr$$

$$= \frac{1}{4\pi} \int_0^{r_{max}} nL(\varepsilon) \, dr = \frac{nL(\varepsilon) r_{max}}{4\pi}$$

where n is the number of sources per unit volume. We can take into account the effects of the Hubble expansion, to a first approximation, by noting that the expansion velocity, v, is given by

$$v = Hr,$$

where H is Hubble's constant, and then using as the limit of our integration the distance at which the expansion velocity becomes equal to half the velocity of light, i.e.

$$r_{max} = c/2H$$

This gives

$$j(\varepsilon) \sim \frac{cnL(\varepsilon)}{8\pi H}.$$

A more detailed calculation (Lichti, Bignami, and Paul 1978) gives

$$j(\varepsilon) = \frac{cn}{4\pi H} \int_0^{z_{max}} \frac{L\{\varepsilon(1 + z)\}}{(1 + z)(1 + 2q_0 z)^{1/2}} \, dz,$$

where $z = \Delta\lambda/\lambda$ and q_0 is the cosmological deceleration parameter.

We see that the contribution of any particular class of sources to the diffuse flux depends on the product of the luminosity times the space density of the sources. The space density of Seyfert galaxies is four orders of magnitude greater than that of quasars but, from our limited knowledge of the gamma ray emission of active galaxies, it seems that the luminosities of Seyfert galaxies may be three orders of magnitude less. Bignami et al. (1979) have shown that both quasars and Seyfert galaxies could make a large contribution to the diffuse flux but there is still a great deal of uncertainty because of the very limited sample of active galaxies from which gamma rays have been detected and because of the possibility of evolutionary effects in these galaxies at very large red shifts.

Stecker (1971) has proposed that the diffuse gamma rays originated in the decay of neutral pions which were produced by matter–antimatter annihilation in the early stages of the universe. A universe which starts with nearly equal amounts of matter and antimatter has a certain theoretical appeal because it reflects the high degree of symmetry between particles and antiparticles which is observed in elementary particle physics. The spectrum of the gamma rays from the decay of neutral pions has a maximum at ~ 70 MeV and Stecker assumes that the annihilation took place at $z \sim 50$–100 so that this feature is now observed at a few MeV.

References

Auriemma, G., Angeloni, L., Belli, B. M., Bernadi, D., Cardini, D., Costa, E., Emanuele, A., Giovannelli, F., and Ubertini, P. (1978). *Astrophys. J.* **221**, L7.

Baker, R. E., Butler, R. C., Dean, A. J., Di Cocco, G., Dipper, N. A., Martin, S. J., Mount, K. E., Ramsden, D., Barbaglia, G., Barbareschi, G., Boella, G., Bussini, A., Igiuni, A., Inzani, P., Perotti, F., and Villa, G. (1979). *Nucl. Instrum. Methods* **158**, 595.

Beall, J. H. (1979). *Ph.D. Thesis*, University of Maryland.

——, Rose, W. K., Graf, W., Price, K. M., Dent, W. A., Hobbs, R. W., Conklin, E. K., Ulich, B. L., Dennis, B. R., Crannell, C. J., Dolan, J. F., Frost, K. J., and Orwig, L. E. (1978). *Astrophys. J.* **219**, 738.

Bignami, G. F., Bennett, K., Buccheri, R., Caraveo, P. A., Hermsen, W., Kanbach, G., Lichti, G. G., Masnou, J. L., Mayer-Hasselwander, H. A., and Paul, J. A. (1981). *Astron. Astrophys.* **93**, 71.

——, Fichtel, C. E., Hartman, R. C., and Thompson, D. J. (1979). *Astrophys. J.* **232**, 649.

Boldt, E. (1981). Comments on Astrophys. **9**, 97.

Bridgeland, M. T. (1972). *Ph.D. Thesis*, University of Bristol.

Cutri, R. M. and Rudi, R. J. (1980). *Astrophys. J.* **241**, L141.

Dean, A. J. and Ramsden, D. (1981). *Phil. Trans. R. Soc. A* **301**, 577.

Dyer, C. S., Engel, A. R., and Quenby, J. J. (1972). *Astrophys. space Sci.* **19**, 359.

Elvis, M., Maccacaro, T., Wilson, A. S., Ward, M. J., Penston, M. V., Fosbury, R. A. E., and Perola, G. C. (1978). *Mon. Not. R. astr. Soc.* **183**, 129.

Fichtel, C. E., Simpson, G. A., and Thompson, D. J. (1978). *Astrophys. J.* **222**, 833.

Grindlay, J. E. (1975). *Astrophys. J.* **199**, 49.

——, Helmken, H. F., Hanbury-Brown, R., Davis, J., and Allen, L. R. (1975). *Astrophys. J.* **197**, L9.

Hall, R. D., Meegan, C. A., Walraven, G. D., Djuth, F. T., and Haymes, R. C. (1976). *Astrophys. J.* **210**, 631.

Jones, T. W., O'Dell, S. L., and Stein, W. A. (1974). *Astrophys. J.* **192**, 261.

Katz, J. (1976). *Astrophys. J.* **206**, 910.

Lichti, G. G., Bignami, G. F. and Paul, J. A. (1978). *Astrophys. space Sci.* **56**, 403.

McBreen, B. (1979). *Astron. Astrophys.* **71**, L19.

Marshall, F. E., Boldt, E. A., Holt, S. S., Miller, R. B., Mushotzky, R. F., Rose, L. A., Rothschild, R. E., and Serlemitsos, P. J. (1980). *Astrophys. J.* **235**, 4.

Meegan, C. A. and Haymes, R. C. (1979). *Astrophys. J.* **233**, 510.

Mushotzky, R. F. (1977). *Nature (London)* **265**, 225.

——, Holt, S. S., and Serlemitsos, P. J. (1978a). *Astrophys. J.* **225**, L115.

——, Serlemitsos, P. J., Becker, R. H., Boldt, E. A., and Holt, S. S. (1978b). *Astrophys. J.* **220**, 790.

Perotti, F., Della Ventura, A., Sechi, G., Villa, G., Di Cocco, G., Baker, R. E., Butler, R. C., Dean, A. J., Martin, S. J., and Ramsden, D. (1979). *Nature (London)* **282**, 484.

——, ——, Villa, G., Di Cocco, G., Bassani, L., Butler, R. C., Carter, J. N. and Dean, A. J. (1981). *Astrophys. J.* **247**, L63.

Primini, F. A., Cooke, B. A., Dobson, C. A., Howe, S. K., Scheepmaker, A., Wheaton, W. A., Lewin, W. H. G., Baity, W. A., Gruber, D. E., Matteson, J. L., and Peterson, L. E. (1979). *Nature (London)* **278**, 234.

Schreier, E. J., Feigelson, E., Delvaille, J., Giacconi, R., Grindlay, J., and Swartz, D. A. (1979). *Astrophys. J.* **234**, L39.

Schlickeiser, R. (1980). *Astrophys. J.* **240**, 636.

Stecker, F. W. (1971). *Cosmic gamma rays*. Mono Book Corporation, Baltimore, MD.

Trombka, J. I., Dyer, C. S., Evans, L. G., Bielefeld, M. J., Seltzer, S. M., and Metzger, A. E. (1977). *Astrophys. J.* **212**, 925.

Tucker, W. H. (1975). *Radiation processes in astrophysics*, M.I.T. Press, Cambridge, MA.

Weedman, D. W. (1977). *Annu. Rev. Astron. Astrophys.* **15**, 69.

White, R. S., Dayton, B., Gibbons, R., Long, J. L., Zanrosso, E. M., and Zych, A. D. (1980). *Nature (London)* **284**, 608.

INDEX